PHYSICS AND CHEMISTRY OF POROUS MEDIA II

AIP CONFERENCE PROCEEDINGS 154

RITA G. LERNER
SERIES EDITOR

PHYSICS AND CHEMISTRY OF POROUS MEDIA II

RIDGEFIELD, CT 1986

EDITORS:
JAYANTH R. BANAVAR, JOEL KOPLIK
& KENNETH W. WINKLER
SCHLUMBERGER-DOLL RESEARCH

AMERICAN INSTITUTE OF PHYSICS   NEW YORK 1987

Authorization to photocopy items for internal or personal use, beyond the free copying permitted under the 1978 US Copyright Law (see statement below), is granted by the American Institute of Physics for users registered with the Copyright Clearance Center (CCC) Transactional Reporting Service, provided that the base fee of $3.00 per copy is paid directly to CCC, 27 Congress St., Salem, MA 01970. For those organizations that have been granted a photocopy license by CCC, a separate system of payment has been arranged. The fee code for users of the Transactional Reporting Service is: 0094-243X/87 $3.00.

Copyright 1987 American Institute of Physics

Individual readers of this volume and non-profit libraries, acting for them, are permitted to make fair use of the material in it, such as copying an article for use in teaching or research. Permission is granted to quote from this volume in scientific work with the customary acknowledgment of the source. To reprint a figure, table or other excerpt requires the consent of one of the original authors and notification to AIP. Republication or systematic or multiple reproduction of any material in this volume is permitted only under license from AIP. Address inquiries to Series Editor, AIP Conference Proceedings, AIP, 335 E. 45th St., New York, NY 10017.

L.C. Catalog Card No. 83-73640
ISBN 0-88318-354-4
DOE CONF-8610148

Printed in the United States of America

CONTENTS

I. Mictostructure

The Morphology of Porous Sedimentary Rocks 3
    M. H. Cohen
A Model for Dolomitization by Pore Fluid Flow 17
    J. R. Wood
T1-Permeability Correlations 37
    R. Dashen, P. Day, W. Kenyon, C. Straley, and J. Willemsen
Origin of Porosity in Synthetic Materials 63
    D. W. Schaefer, J. P. Wilcoxon, K. D. Keefer, B. C. Bunker, R. K. Pearson, I. M. Thomas, and D. E. Miller

II. Fluid Flow

Incorporating the Influence of Wettability into Models of Immiscible Fluid Displacement Through a Porous Media 83
    E. B. Dussan V.
Statistical Physics and Immiscible Displacements Through Porous Media 98
    R. Lenormand
Stable and Unstable Miscible Flows Through Porous Media 116
    J.-C. Bacri, N. Rokotomala, and D. Salin
The Transport Properties of Non-dilute Suspensions. Renormalization via an Effective Continuum Method 129
    A. Acrivos and E. Chang

III. Chemistry

Progressive Chemical Modification of Clastic Sediments with Burial 145
    C. D. Curtis
The Major Ion Chemistry of Saline Brines in Sedimentary Basins 160
    L. S. Land
Energetics of Complex Aluminosilicates 180
    A. Navrotsky
Experimental Studies on Physical Properties of Vermiculite, Muscovite, and Kaolinite Clays 195
    N. Wada

IV. Mechanical Properties

Extensions of Biot's Theory of Poroelasticity to Complex Porous Media 209
    J. G. Berryman and L. Thigpen
Porosity and the Brittle-Ductile Transition in Sedimentary Rocks 229
    J. M. Logan
Probing Porous Media with 1st Sound, 2nd Sound, 4th Sound, and 3rd Sound 243
    D. L. Johnson, T. J. Plona, and H. Kojima

V. Electrical Properties

Electrical Properties from 10^{-3} to 10^{+9} Hz: Physics and Chemistry 281
    G. R. Olhoeft

Percolation in Oil-Continuous Microemulsions 299
    S. Bhattacharya, J. P. Stokes, M. J. Higgins, M. W. Kim, and J. S. Huang

Fractal Surfaces in Porous Media 304
    P.-z. Wong

Foreword

The burgeoning scientific interest in the structural and transport properties of porous media has produced a flood of recent research. As the subject is an unusually interdisciplinary one, conferences provide the ideal setting to compare, evaluate, and summarize recent work. This volume contains the proceedings of the Second International Symposium on the Physics and Chemistry of Porous Media, held at Schlumberger-Doll Research in Ridgefield, Connecticut on October 15–17, 1986. The meeting featured 20 invited talks, of which 18 are represented in this volume, and was attended by about 140 scientists. The papers have been organized, somewhat arbitrarily, under the headings of microstructure, fluid flow, chemistry, mechanical properties and electrical properties.

We are indebted to Fred Gamble and M. Stanley Whittingham for their consistent support and encouragement. Cathy Green and Jeanette Gerfin efficiently handled the organizational details. We would also like to thank the session chairmen—Bill Murphy, Mike King, Dick Plumb, Steve Cramer and Bob Guyer.

Jayanth R. Banavar
Joel Koplik
Kenneth W. Winkler

Ridgefield, Connecticut
December, 1986

I. Microstructure

THE MORPHOLOGY OF POROUS SEDIMENTARY ROCKS

Morrel H. Cohen

Exxon Research and Engineering Company
Route 22 East
Annandale, New Jersey 08801

ABSTRACT

The salient morphological features of porous sedimentary rocks are (1) absence of significant grain growth, (2) persistence of porosity, (3) continuity of the pore space down to low values of the porosity, (4) sheet-like pores at low porosity, (5) small intergranular contacts, and (6) fractal or at least rough pore-grain interfaces. These are astonishing when one recognizes that there had been ample time for sintering to occur. The first four features follow from the fifth, small contacts, which can be understood if the pore-grain interfacial free energy is less than one-half the grain-boundary energy, as is the case for pure silica and probably for other typical rock constituents. Sintering is suppressed, and, if the interfacial free energy is low enough, the interface is rough, feature (6).

INTRODUCTION

The formation of sedimentary rocks in sedimentary basins can be described as follows. Rock grains - formed by erosion, sand, dust, or skeletons of marine animals - fall to the bottom of a body of water overlying the sedimentary basin. This process of sedimentation buries earlier sediments which descend downward into the earth as the process continues. As burial and descent proceeds, the rock grains are compacted into aggregates with an initial porosity ϕ_o, the volume fraction occupied by the pores between the grains. As the sediments descend further, the temperature and pressure increases, chemical reactions can take place, and, over long periods of time, the aggregates are transformed into rocks, sedimentary rocks with porosity $\phi < \phi_o$.

We are concerned in this paper with understanding the salient morphological features of the sedimentary rocks.[1] We shall see that these features are astonishing if one supposes that the process of rock formation involves sintering, as sintering is conventionally understood.[2] We shall propose instead that the unexpected morphological features derive from a low and even negative free energy of the interface between the rock grains and the water within the pores; that is, that the dominant morphological features observed in sedimentary rocks are controlled by interfacial-free-energy relations.

SINTERING AND THE EXPECTED MORPHOLOGY

Deeply buried sediments could have been in place under the earth for as long as 10^8 years. There they would have experienced temperatures as high as 200°C and pressures as high as 1000 atm. Under those circumstances, the dominant kinetic pathway would be dissolution of solid material from the grains, diffusion and/or convection within the pore space, and reprecipitation at the grain-pore interfaces. The rate-limiting step along that pathway is dissolution. As long as the activation energy for dissolution is less than about 3 eV, there is ample time over the life of the rock for the system to have started moving towards equilibrium, at first locally and then globally.

We can regard the initial state of the sedimentary rock as that of the earliest compaction. The rock grains are constrained by their neighbors, but the porosity is high, about 40%, and the contacts between the grains are very small. The pore space is filled by water containing dissolved minerals, an electrolyte solution.

We are familiar with the morphological development of other such granular materials when sintered in air or in inert atmospheres. The first step is the establishment of local equilibrium under the influence of surface tension, as indicated in Fig. 1a. Consider two grains in contact at a grain boundary.

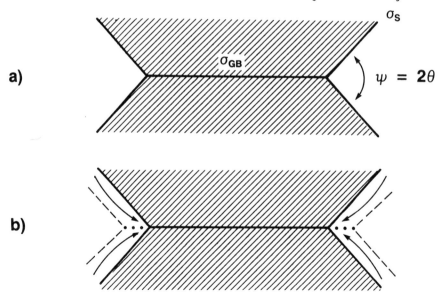

Figure 1. a. Local equilibrium under the forces of surface and grain boundary tensions. b. Transport of material to the contact during sintering.

The outer grain surfaces subtend a dihedral angle $\psi = 2\phi$ at the boundary with ϕ given by

$$\sigma_{GB} = 2\sigma_S \cos\phi \qquad (1a)$$

$$x \equiv \frac{\sigma_{GB}}{2\sigma_S} < 1, \qquad (1b)$$

where σ_S is the interfacial free energy at the external surface of the grain and σ_{GB} is the grain-boundary energy.

As annealing proceeds, sintering occurs, but on a longer time scale.[2] As indicated in Fig. 1b, material is transported to the contact areas, and the contacts grow. In this process external surface area is destroyed and replaced by grain boundary in response to the driving force

$$2\sigma_S - \sigma_{GB} > 0 . \qquad (1c)$$

As a consequence, the porosity decreases even in the absence of external pressure; it rounds up, it disconnects for $x \leq 0.83$, and it disappears completely for $x \leq 0.5$. The next and still slower stage is the growth of larger grains at the expense of smaller grains, a coarsening or ripening of the structure which reduces the grain boundary area in response to the driving force $\sigma_{GB} > 0$. Ultimately, there is a disconnection and disappearance of the porosity and a disappearance of the granularity. The porous compact of fine grains becomes a solid mass of large grains as the sintering reaches its final stages.

THE ACTUAL MORPHOLOGY

The actual morphology of certain typical classes of sedimentary rocks[1] is nothing like the morphology of typical sinter bodies as described above. In what follows, we shall list the six features we regard as both most significant and most typical. We have winnowed these from a vast body of information derived from microscopic examination by others of rock specimens and their pore casts as well as from the implications of various physical properties of rocks for their internal structure as inferred by us from recently developed theories. In doing so, we have deliberately ignored what we regard is inessential detail or inconsequent variability.

1. There is little evidence for significant grain growth.

2. Porosity persists as long as the sediments were not buried too deeply or exposed to igneous intrusions, $\phi \neq 0$.

3. The pore space remains connected down to very low values of ϕ,[3] values much lower than the 3% expected for the percolation threshold consequent to contact growth.[4]

4. As ϕ decreases, the pores become sheet-like instead of rounding up.[5,6]

5. The intergranular contacts remain small, that is

$$c \ll G, \qquad (2)$$

where G is a typical grain diameter and c is a typical contact diameter. Moreover, there is evidence from limestones in which residual cellular structure provides markers that material moves <u>away from</u> instead of <u>into</u> the contacts.[7]

6. It has been found that the pore-grain interface is commonly rough and inferred that the interface is fractal and the pore space is fractal as well, with the same fractal dimensions D_s and D_v,[8,9,10]

$$D_s = D_v. \qquad (3)$$

The first five features are diametrically opposite to the corresponding features of sinter bodies. The persistence of porosity and especially its continuity, features 2 and 3, provide the physical basis for the oil and gas industries and for the hydrology of the earth. Were sintering to have taken place, there would be no large scale underground water movements nor would there be an oil or gas industry as we know it today.

SOME TOPOLOGY

Before considering the implications of the above morphological features, we shall need briefly to review the relevant topology.[11,1] The pore space can be mapped into a graph, or skeleton, while rigorously preserving the topology. The corresponding skeleton of the grain space is a graph which interpenetrates that of the pore space and is, in a certain sense, dual to it. These mappings can be carried out in such a way as to partition the entire space, pore space plus grain space, into space-filling cells each containing one rigorously defined grain. The faces of these cells are coincident with the grain boundaries as they pass through the contact areas. However, they not only separate the grains at the contacts, they also partition the pore space. The edges and vertices of these cells comprise the graph which is the skeleton of the pore space. The pore space in the neighorhood of each vertex can be regarded as a pore chamber; that in the neighborhood of each edge can be regarded as a pore channel.

SMALL CONTACTS ARE THE KEY

Consider Fig. 2 in which two grains and the neighboring pore space are shown partitioned by a cell face. Suppose that the contacts are constrained to remain small through the operation of some as yet unspecified mechanism, as in feature 5 of

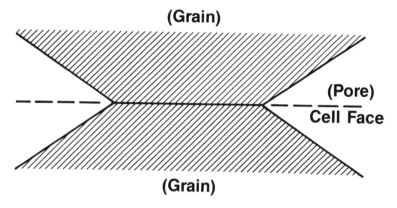

Figure 2. Partitioning of two grains in contact and the neighboring pore space by a cell face.

Section III. Then granularity must persist. Grains remain sharply defined, and the normal mechanism for grain growth is severely inhibited. Thus feature 1 is a consequence of the persistence of small contact areas. Similarly porosity must persist, because small contact areas imply the continued existence of the pore-grain interface, feature 2. The pore space must remain continuous, because the discontinuity occurs as pore channels are closed by contact area growth. This defines a percolation problem with a threshold value for the ratio c/G which is of order 1/4, much larger than occurs in sedimentary rocks. Finally, if the contact areas are small, the pores are constrained in two dimensions. The only way the porosity can decrease is by collapse of the pore space towards the partitioning surface made up of the cell faces, as in Fig. 2. This leads immediately to a sheet-pore morphology, a connected network of sheets, as indicated in Fig. 3.

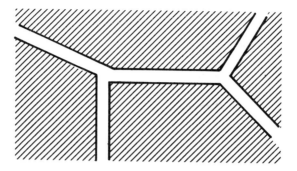

Figure 3. The sheet-pore morphology.

Thus, the first four morphological features follow directly from the persistence of small contacts. Our problem becomes one of finding the mechanism by which this occurs.

THE ANTISINTERING HYPOTHESIS

If the contacts remain small, there can be no driving force causing their growth. That is, the inequalities in Eqs. (1b) and (1c) must be violated, i.e.

$$x = \frac{\sigma_{GB}}{2\sigma_{GL}} > 1 \qquad (4a)$$

$$2\sigma_{GL} - \sigma_{GB} > 0, \qquad (4b)$$

where σ_{GL} is the interfacial free energy between the grains and the pore fluid. Let us therefore make the bold hypothesis that Eq. 4 is normal for sedimentary rocks, that for sedimentary rocks an <u>antisintering regime</u> characterized by Eq. 4 holds, instead of the more familiar sintering regime characterized by Eq. 1.

ITS CONSEQUENCES

We can examine the consequences for contact size of the antisintering hypothesis with the help of the Johnson, et al. theory of contacts.[12] They derive from macroscopic elasticity theory the following expression for the radius $\underline{a}$ of the contact between two elastic spheres of Young's modulus, E, Poisson ratio ν, and radius R under a load P:

$$\underline{a}^3 = \frac{3}{4} \frac{1-\nu^2}{E} R \left[P + 3\pi R\gamma + \{(3\pi R\gamma)^2 + 6\pi R\gamma P\}^{1/2} \right]. \qquad (5)$$

Here γ is the free-energy required to separate unit area of the contact between two spheres.

Eq.(5) can be applied to rock grains in contact with their interstitial pore space permeated by pore water provided the contact size is small compared to contact separation and grain size, which is indeed the case. ν, E, and R are then to be interpreted as mean or typical values, and γ is given by

$$\gamma = 2\sigma_{GL} - \sigma_{GB} . \qquad (6)$$

The load P is given by

$$P = S(P_R - P_W)/Z , \qquad (7)$$

where S is the rock-water interfacial area per grain, Z is the number of intergranular contacts per grain, P_R is the pressure within the rock grains or rock framework, and P_W is the pressure within the pore fluid.

Now γ is negative in the antisintering regime, and (5) can be rewritten as

$$\underline{a}^3 = \frac{3}{4}\frac{1-\nu^2}{E} R [P - P_0 + \{P_0^2 - 2P_0 P\}^{1/2}], \qquad (8)$$

where

$$P_0 = 3\pi R|\gamma|. \qquad (9)$$

Inspection of (8) shows that there is no value of P for which the quantity in square brackets is both real and positive. Thus, there can be no macroscopically large value of $\underline{a}$ for a contact in local equilibrium under elastic and surface-tension forces in the antisintering regime, according to this theory. A nonequilibrium contact of macroscopic size must therefore shrink until it reaches equilibrium at a size $\underline{a}$ below the lower limit of the size regime in which macroscopic elasticity theory and the concept of surface free energy are adequate, a few tens of Angstroms.

The Johnson, et al. theory is incomplete, taking into account mechanical effects only and not allowing for rearrangement of the interfaces under the influence of the forces of surface tension via material transport. Correcting this omission, however, would only strengthen the conclusion of the last paragraph.

As implied above, there are two classes of mechanisms by which the internal morphology of the rock changes in response to increased differential pressure between the pore fluid and the rock framework, P_R-P_W, mechanical adjustment and matter transport with the former much faster than the latter. Both are very much faster than the rate of pressure change, in general, and the local morphology is in secular equilibrium under the local pressure, temperature, and chemical conditions. Nevertheless, one can imagine the morphological response to a secular increase in $P_R - P_W$ from zero to its actual value as occurring in two steps, the first being entirely mechanical and the second consisting of matter transport with concomitant mechanical adjustments. The first step would give rise to an increase in contact radius from zero to

$$\underline{a}^3 = \frac{3}{4} \frac{1 - \nu^2}{E} R P \qquad (10)$$

with P given by (7). Associated with this contact formation would be a displacement δ of the grain interior towards the contact given by[13]

$$\delta^3 = \frac{9}{2} (\frac{1 - \nu^2}{E})^2 \frac{P^2}{R} \qquad (11)$$

That is, matter flows from the interior of the grain into the intergranular contact area. During the second step, the contact radius is reduced to its proper microscopic value by dissolution of material in the vicinity of the Newton-Hertz contact formed in the first step, thus reducing both surface free energy and strain energy. That is, matter flows from the contact area into the pore fluid from which it precipitates on the rock-water interface away from the contacts. This flow of matter from the rock interior to the contacts and from the contacts through the pore fluid to distant regions of the rock-fluid interface is sketched in Fig. 4.

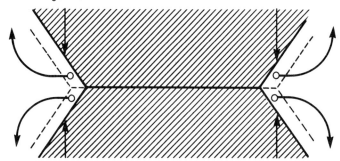

Figure 4. Flow of material from the grain interior to the grain boundaries, across the grain-pore interfaces into the pore fluid, and through the fluid to the interface away from the contacts during compaction in the antisintering regime.

In this way, the forces of surface tension maintain small contacts in the antisintering regime, automatically yielding the granularity, porosity, continuity of porosity, and sheet pore morphology observed in sedimentary rocks and so hard to understand if the sintering regime were to obtain.

IS IT POSSIBLE?

Can the interfacial free energy of typical rock materials actually fall below one-half the grain boundary energy as is required for the antisintering regime? That question can be answered for the simplest of all silicates, pure fused silica.[14] In fused silica, the silicon atoms are tetrahedrally coordinated with each other through bridging oxygen atoms. Silicon atoms at the surface will be three-fold coordinated and oxygen atoms there only one fold coordinated. The structure can of course relax to compensate partially for these coordination deficiencies, but the net result is that fused silica has a large surface energy, 260 ergs/cm^2.

When the surface is exposed to water and monolayer coverage achieved, each surface silicon atom becomes four-fold coordinated by the addition of an OH, and each surface oxygen atom becomes two-fold coordinated by the addition of an H-atom. The surface is completely decorated with OH, and the silica becomes a giant polysilanol molecule. The surface energy is thereby reduced by a factor of two to 130 ergs/cm^2.

When the surface is then covered by bulk water a heat of wetting of 190 ergs/cm^2 is released. Thus the interfacial free energy of silica grains in water is in fact negative; σ_{GL} is -60 ergs/cm^2. So, with silica grains, we are not only in the antisintering regime, we are all the way into the domain of negative surface energies, and new phenomena are possible. One can anticipate this to be the case generally among silicates, alumino-silicates and other common constituents of rocks.

FRACTAL INTERFACES

When the microstructure of rocks is examined at relatively low resolution, that is via optical microscopy, the pore-grain interfaces appear relatively smooth. When higher resolution is used, via electron microscopy, the pore-grain interface shows features at all length scales between an estimated lower limit L_1 of ~20 Å and an upper limit L_2 of order the grain size. The frequency distribution of feature size falls off as a power of the feature size between these limits, demonstrating that the interface is fractal on length scales between L_1 and L_2. Katz and Thompson, who obtained these beautiful results,[8] then went on to estimate the porosity on the assumption that the pore space as well was fractal with the same fractal dimension, D_v, as the surface, D_s, $D_v = D_s$. They found agreement within experimental error between measured porosities and porosities predicted on the assumption of a fractal pore space with

$D_V = D_S$. Apparently, not all sedimentary rocks are fractal, but fractal rocks with $D_V = D_S$ are common, morphological feature 6 of Section III.

The particular material reported in most detail by Katz and Thompson was a clean sandstone.[8] Their technique did not distinguish between the two possible types of fractal, self-affine and self-similar.[15] X-ray scattering, however, does permit making such a distinction, and the results of Wong indicate that clean sandstones have self-affine interfaces, whereas in clay-containing limestones at least that portion of the interface associated with the clay appears to be self-similar.[10]

Consider first the case of a clean sandstone and self-affine fractals. Given that all other morphological features require at least a low positive value of σ_{GL} and that we have shown that even negative values of σ_{GL} can occur, we should ask what the consequences of low or negative surface energy are for the surface configuration. At low temperatures, a rough interface clearly has higher free energy than a smooth interface for $\sigma_{GL} > 0$. However, as the temperature increases, the additional entropy associated with a rough surface lowers the free energy of the rough relative to the smooth surface until at a temperature T_R the free energy of the rough surface overtakes that of the smooth and the surface spontaneously roughens.[16] This roughening transition at the roughening temperature T_R can also be viewed as occurring at fixed temperature as σ_{GL} is varied through a critical value $\sigma_R > 0$. The rough interface is a self-affine fractal between a lower limit L_1 of microscopic size set by curvature effects and an upper limit L_2 of order the grain size or smaller.[17] The amplitudes of the excursions of the surface increase as σ_{GL} decreases further below σ_R. Ultimately, the excursions are limited by encounters with the roughened surfaces of neighboring grains. At that point, the entire pore space is filled by the interface and has become fractal as well. Since the typical separation between bits of surface has by then become L_1, the fractal dimension of the pore volume D_V must equal that of the pore surface, D_S. The surface to volume ratio of the pore space is then L_1^{-1}.

In contrast to the well-defined value of interfacial free energy σ_R at which the roughening transition occurs, the value of σ at which the pore volume becomes fractal is not well defined, depending on grain size and porosity. However, it must in any case be nonnegative. Thus, if we are dealing with rocks having negative σ_{GL}, we can be sure that both their pore surface and pore volume are fractal with $D_V = D_S$.

This mechanism of generation of fractal structures via spontaneous surface roughening is probably the sole mechanism for clean, clay-free sediments. When clay is present as well, another mechanism occurs. Beginning with the earliest stages of sedimentation, the clay particles, very much smaller than the other rock grains present, will find themselves within a pore space defined by the larger rock grains. Within that pore space, the clay particles can diffuse, but only very slowly. Thus, one has the possibility of diffusion-limited aggregation[19] of the clay particles within the pore space. Diffusion-limited aggregation leads to self-similar fractals.[19] If the aggregation proceeds to completion, the morphology of the pore space will consist of regions empty of clay particles and regions containing clay in the most tightly packed self-similar fractal structure possible. Because of the affinity between clay and water and the polar layer structure of the clay, the water-filled pore space permeates the clay aggregate but with $D_S = D_V = 3$, as is observed,[16,20] another consequence of negative σ_{SL}.

SUMMARY AND CONCLUSIONS

We have identified as the key morphological features of sedimentary rocks (1) granularity, (2) porosity, (3) continuity of the pore space, (4) the development of sheet pores as the porosity becomes small, (5) small contacts, and (6) fractal interfaces and pore spaces with $D_S = D_V$. These features would be astonishing if conditions favoring sintering existed, i.e. if $x = \sigma_{GB}/2\sigma_{GL} < 1$. Accordingly, we have supposed that $x = \sigma_{GB}/2\sigma_{GL} > 1$ occurs and sedimentary rocks are in the anti-sintering regime. Feature 5, small contacts, immediately follows, and, as a consequence, features 1 to 4 follow as well. We have also pointed out that σ_{GL} is negative for water-wet silica, showing that the requirement that σ_{GL} be less than $(1/2)\sigma_{GB}$ is not extreme and suggesting an explanation of feature 6 as spontaneous surface roughening. We can therefore conclude that σ_{GL} may well control the morphology of sedimentary rocks and even that $\sigma_{GL} < 0$ may be common. The various morphological regimes we have identified are summarized in Fig. 5.

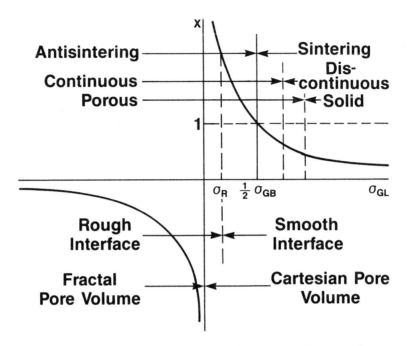

Figure 5. Summary of the various morphological regimes.

Having placed so much emphasis on the low interfacial free energies of water-wet rocks, we must consider what happens when the water is displaced by other fluids, namely gas and oil. The case of gas is particularly troublesome because we can be sure that x < 1 for pure methane. However, one must recognize that one is dealing with a three phase system, and, as long as the spreading pressure

$$\pi = \sigma_{GI} - \sigma_{Gw} - \sigma_{wI} \qquad (5)$$

is positive, where I indicates the invading fluid, the rocks will remain water wet by a film of water interposed between the grains and the invading fluid. Further if $\sigma_{Gw} + \sigma_{wI}$ remains below $1/2\ \sigma_{GB}$, the rock will remain in the antisintering regime. Clearly, because of the additional surface energy σ_{wI}, fractal structures are less likely but not impossible. Finally, if σ_{GI} is sufficiently low, as it may be for some oil and rock pairs, that π is negative, it is highly likely that the rocks will still be in the antisintering regime.

We described in Section III the six salient morphological features of certain typical classes of sedimentary rocks. We have explained these by supposing that for those classes of sedimentary rocks their morphology is controlled by the interfacial free energies and, in particular, that σ_{GL} is low relative to σ_{GB} and possibly even negative. However, not all sedimentary rocks show all of these features. One does not always solely see sheet pores at low porosities. One does not always see small contacts, as, for example, in heavily cemented rocks. In general, when these six key features are not observed, it is clearly implied that the conditions for their manifestation were not present during diagenesis. Their absence in some rocks does not constitute a contradiction of the present considerations, but rather a confirmation of the sensitivity of morphology to such variables as the interfacial free energies.

REFERENCES

1. M. H. Cohen and M. P. Anderson, The Electrochem. Soc., 133 (1986).
2. G. C. Kuczynski, Sintering Processes, Plenum Press, New York (1980).
3. P. N. Sen, C. Scala and M. H. Cohen, Geophysics, 46, 781 (1981) and references therein.
4. J. N. Roberts and L. M. Schwartz, Phys. Rev. B, 31, 5990 (1985).
5. E. Pittman, Physics and Chemistry of Porous Media, eds. D. L. Johnson and P. N. Sen, AIP Conference Proceedings, No. 107, 1984, p. 1.
6. N. C. Wardlaw, Am. Assoc. Petroleum Geologists Bull., 60, 245 (1986); N. C. Wardlaw and J. P. Cassan, Bull. Can. Petroleum Geology, 26, 572 (1978); 27, 117 (1979). N. C. Wardlaw, A. Oldershaw and Mavis Stout, Can. J. Earth Sci., 15, 1861 (1978).
7. N. C. Wardlaw, private communcation.
8. A. J. Katz and A. H. Thompson, Phys. Rev. Lett., 54, 1325 (1985).
9. P. I. Hall, D. F. R. Milner and R. L. Borst, J. Electrochem Soc., 133 (1986).
10. P.-z. Wong, Phys. Rev. B 32, 7417 (1985); P.-z. Wong, J. Howard, and J.-S. Lin, Phys. Rev. Lett. 57, 637 (1986).
11. M. H. Cohen and C. Lin, "Macroscopic Properties of Disordered Media", eds. R. Burridge, S. Childress and G. papanicolan, Springer Verlag (1981) p. 74; C. Lin and M. H. Cohen, J. Appl. Phys., 53, 4153 (1982).
12. K. L. Johnson, K. Kendall and A. D. Roberts, Proc. Roy. Soc. Lond., A334, 301 (1971).
13. H. Hertz, Miscellaneous Papers, MacMillan, London, 1986, p. 146.
14. R. K. Iler, "The Chemistry of Silica", Wiley-Interscience, 1979, p. 645.

15. B. B. Mandelbrot, "The Fractal Geometry of Nature", Freeman, 1982; R. F. Voss, NATO ASI Series, Vol. F17, "Fundamental Algorithms for Computer Graphics", ed. R. A. Earnshaw, Springer-Verlag, Berlin Heidelberg, 1985.
16. W. K. Burton and N. Cabrera, Discuss. Faraday Soc. $\underline{5}$, 33 (1949).
17. S. Alexander, preprint.
18. T. Witten, "On Growth and Form", eds. H. E. Stanley and N. Ostrowsky, Martinius Nijhoff, Publishers, 1986, p. 54.
19. This argument is based on scattering results of Ball and Sinha and geometric considerations by Sinha. I am grateful to Dr. Sinha for communicating them to me in advance of publication.

A MODEL FOR DOLOMITIZATION BY PORE FLUID FLOW

J. R. Wood, COFRC, La Habra, CA 90631

INTRODUCTION

Purpose of Paper

The purpose of this paper is to describe a model for the dolomitization of massive carbonates in which the principal diagenetic processes is pore fluid flow. This model is the outgrowth of similar modeling efforts (e.g., Wood and Hewett, 1982, 1984, 1986; Davis et al. 1985; Hewett, 1986; in which the rock alteration is assumed to occur when saturated pore fluids migrate across a temperature field. This process will drive diagenetic reactions if the pore fluids are assumed to maintain chemical equilibrium with the host rock because all minerals have at least a small temperature dependence and a small amount of solid must continuously precipitate or dissolve as the pore fluid encounters local temperature changes.

This paper will concentrate on a single specific diagenetic event, carbonate dolomitization and will attempt to account for the development and spatial distribution of the dolomite in terms of a simple flow model. This particular application will be developed in terms of quantitative description of the various fluid flow and temperature fields involved. The details of the chemical interactions will be deferred.

THE PROBLEM

Conceptual Model For Dolomitization

The conceptual geologic model for this study is depicted in Figure 1. For simplicity we will assume that a carbonate build-up sits directly on an impermeable granitic basement.

Similarly the carbonate unit will be bounded on the top by semi-permeable rocks, salts and/or shale, which will guide the flow but which will be sufficiently "leaky" to permit fluids to pass out of the carbonate basinward. In this model, the upper and lower lithologic boundaries will function as streamlines.

Generalized Flow Field

These constraints will result in a flow field through the carbonate unit that will be approximately as shown in Figure 1. Here the pore fluid flows down a hydraulic gradient from a recharge area above the intertidal zone, mixes with seawater as it percolates through the carbonate unit and finally exits by transmission across faults or permeability streaks in the upper unit.

We will show below that magnesium-bearing fluids following this flow field can produce a "pod" or mound of dolomite in the original calcitic unit. The dolomite will replace calcite and will be surrounded by unaltered calcite, except at the top where it contacts the upper seal. In particular this model will predict the occurrence of an unaltered layer of calcite <u>below</u> the dolomite.

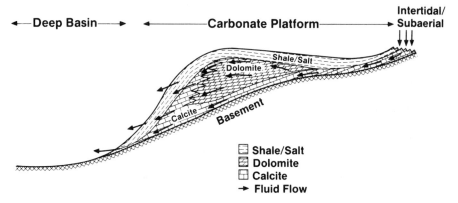

Figure 1 - Conceptual model for dolomitization. Arrows indicate direction of pore fluid flow assuming that meteoric water has access to the formation locally at outcrops, and that it can mix with seawater before penetrating deeply into the formation. The source of the magnesium is assumed to be seawater and the drive for the fluid movement is topographic elevation.

THEORETICAL MODEL FOR DIAGENESIS

The Diagenetic Equation

The basic theory for pore fluid diagenesis has been described (see Wood and Hewett, 1986, for example) and the general equation describing the rate of change of porosity with time due to mineral precipitation and dissolution is:

$$\frac{d\phi}{dt} = \frac{\rho_f}{\rho_s} \alpha_T \bar{V} \cdot \nabla T \qquad (1)$$

where ϕ is porosity, t is time, ρ_f is the fluid density, ρ_s is the solid density, α_T is the thermal mass transfer coefficient (see Wood, 1986, for a discussion of this term), $\underline{V}$ is the fluid flow field, ∇ is the grad operator and T is the temperature field.

For a multimineralic system α_T is an array of coefficients, with different values for each mineral. In the case of a two mineral system, such as calcite-dolomite, this term would be written:

$$\alpha_T = \frac{\alpha_{T(cc)}}{\rho_{cc}} + \frac{\alpha_{T(dol)}}{\rho_{dol}} \qquad (2)$$

where the subscripts cc refer to calcite and dol refer to dolomite. The α_T terms serve to scale the change in porosity for a particular mineral for a given flow field and temperature field.

Accordingly the diagenetic equation can be regarded as the sum of two terms, one for calcite,

$$\frac{d\phi_{cc}}{dt} = \rho_f \frac{\alpha_{T(cc)}}{\rho_{cc}} \underline{V} \cdot \text{grad } T \qquad (3)$$

and one for dolomite,

$$\frac{d\phi_{dol}}{dt} = \rho_f \frac{\alpha_{T(dol)}}{\rho_{dol}} \underline{V} \cdot \text{grad } T. \qquad (4)$$

The overall change in porosity is the sum of these two terms, but expression (3) or (4) by itself describes the change in porosity due to a particular mineral dissolving or precipitating. We can also regard these equations as describing the diagenetic fields for calcite and dolomite respectively.

Equations (3) and (4) can be used to map out the change in porosity due to the precipitation or dissolution of each mineral, although it is only the sum that is generally observable. The $\underline{V} \cdot$ grad T is common to both expressions, and describes the spatial distribution of the alteration. Hence the geometry of the alteration can be calculated just from a consideration of that term, and this paper will be primarily concerned with the $\underline{V}$ grad T term. The next sections describe the temperature, flow, and diagenetic fields used in this model.

The Temperature Field

In most sedimentary basins the temperature field is ordered and oriented in the subsurface as shown in Figure 2. In a typical petroleum basin the temperature increases at some fairly constant rate with depth, usually about 25°C/Km. More complex temperature fields obviously exist, but one characterized by a constant geothermal gradient is the simplest and has the advantage of being the most widely applicable, particularly in petroliferous basins. More complex temperature fields can generally be regarded as perturbations of this field. In almost all cases of diagenetic interest it is reasonable to assume that a linear vertical gradient is a major component of the temperature field.

$$T = T(x, y, z)$$

$$\nabla T = T_x \bar{i} + T_y \bar{j} + T_z \bar{k}$$

$$T_x = \frac{\partial T}{\partial x} \qquad T_y = \frac{\partial T}{\partial y} \qquad T_z = \frac{\partial T}{\partial z}$$

Horizontal Temperature Field

$$\nabla T = T_z \bar{k}$$

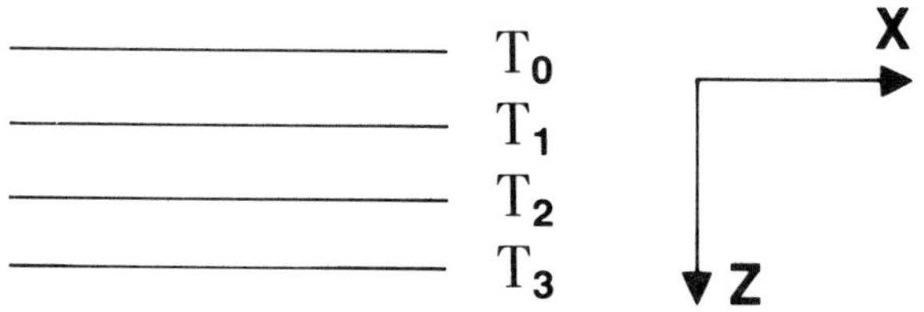

T_0, T_1, T_2, T_3

Figure 2 - Schematic of the temperature field used in the model.

In mathematical terms we have

$$T = T(x,y,z) \tag{5}$$

where T is temperature (C), x and y represent horizontal coordinates and z is the vertical coordinate, and

$$\text{grad } T = T_x \underline{i} + T_y \underline{j} + T_z \underline{k} \tag{6}$$

where T_x, T_y and T_z are the x, y, and z derivatives of T, and $\underline{i}$, $\underline{j}$, and $\underline{k}$ are unit vectors in the x, y, and z directions.

For a horizontal temperature field we have

$$T_x = T_y = 0 \tag{7}$$

and

$$T_z = \text{const.} \tag{8}$$

This is the temperature field used here and is depicted in Figure 2. The isotherms are horizontal planes equally spaced in the z direction.

The Velocity Field

For a two-dimensional flow field we write a generalized stream function as

$$\psi = F(x,z) \tag{9}$$

and denote the components of the flow field as

$$u = -\frac{\partial \psi}{\partial z} \tag{10}$$

and

$$w = \frac{\partial \psi}{\partial x} \tag{11}$$

where u is the <u>horizontal</u> or x component of the flow and w is the <u>vertical</u> or z component. The flow field can then be written

$$\underline{V} = u\,\underline{i} + w\,\underline{k}. \tag{12}$$

We show later how to obtain analytical expressions for u and w from the boundary conditions.

The Diagenetic Field

First, however, we will express the diagenetic field in terms of the fluid flow and temperature fields.

Carrying out the indicated operations on $\underline{V}$ grad T using Equations 6 and 12 above we obtain

$$\underline{V} \cdot \underline{\text{grad}}\ T = w\,T_z. \tag{13}$$

The result simply says that the change in the diagenetic field as a function of fluid flow is proportional to the product of the <u>vertical</u> component of the flow field and the geothermal gradient. Diagenetic alteration can be predicted if we know how the vertical flow component behaves in a given volume of rock.

Our strategy from here on then will be to characterize the fluid flow field, particularly the vertical component. It will turn out that it is the behavior of w(x) that controls the spatial distribution of the dolomitization is concerned.

It should be kept in mind, however, that the simple relationship given by (13) is largely due to two assumptions: (1) the temperature field is horizontal, and (2), a 2-dimensional stream function can adequately characterize the pore fluid flow. Both of these assumptions are open to question, but the position taken here is that this simplified treatment will reveal the general behavior of the system and that more refined treatments can be obtained by analysis of perturbations relative to the simple model.

FORCED FLOW

Stream Function and Velocity Components

The results up to this point have shown that it is the flow field describing the movement of the pore fluid that is of primary importance in developing the diagenetic field. We will concentrate from here on in developing a model for the flow field and then analyzing the diagenetic implications. We begin by sketching out a generalized model (Figure 3).

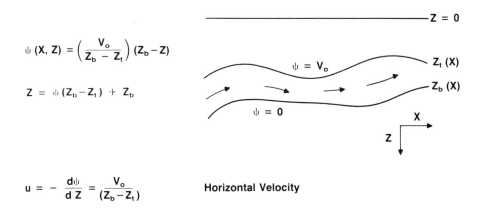

$$\psi(X, Z) = \left(\frac{V_o}{Z_b - Z_t}\right)(Z_b - Z)$$

$$Z = \psi(Z_b - Z_t) + Z_b$$

$$u = -\frac{d\psi}{dZ} = \frac{V_o}{(Z_b - Z_t)} \quad \text{Horizontal Velocity}$$

$$w = \frac{d\psi}{dX} = \frac{V_o}{(Z_b - Z_t)}\left[Z_b' + \left(\frac{Z_b - Z}{Z_b - Z_t}\right)(Z_t' - Z_b')\right] \quad \text{Vertical Velocity}$$

Figure 3 - Sketch of geometry and definitions for streamlines in general case of forced flow. The stream function, ψ, is defined in terms of the functions describing the geometry of the strata, $Z_t(x)$ and $Z_b(x)$ and an arbitrary scale velocity, V_o.

We consider that the carbonate mass to be dolomitized lies between two boundaries which can be taken to be flow boundaries. The upper boundary will be denoted $Z_t(x)$ and the lower boundary will be labeled $Z_b(x)$, where the x in parentheses indicates that these curves are functions of x. These boundaries will lie at some depth below the surface and can have arbitrary shapes with the provision that they vary "slowly" in the x direction. (At this time we exclude abrupt changes in the boundaries which would be typical of faults.)

We will deal only with <u>forced</u>, as opposed to convective, fluid flow, but will not be particularly concerned with the details of the driving forces for the flow.

Let V_o represent the pore fluid velocity. This could be an average velocity across some arbitrary cross section, but a better interpretation can be obtained after defining the stream function and deriving the flow components.

We arbitrarily define a stream function[1] as

$$\psi(x,z) = V_o(Z_b - z)/(Z_b - Z_t) \tag{14}$$

and assume that the flow components calculated from it

$$u = V_o/(Z_b - Z_t) \tag{15}$$

and

$$w = V_o[Z'_b + (Z_b - z)(Z'_t - Z'_b)/(Z_b - Z_t)]/(Z_b - Z_t) \tag{16}$$

where primed quantities indicate differentiation with respect to x, describe the flow field for the pore fluid.

For this choice of stream function, the horizontal component of the flow is quite simple, but it has the desired property that the flow velocity varies inversely with the thickness of the unit. In particular, for a constant thickness, the horizontal flow is constant and V_o can be interpreted as the fluid velocity in a horizontal layer of unit thickness.

The vertical flow field is more complex and involves derivatives of the boundary functions as well as the boundary functions. As might be expected, this complexity is reflected in the diagenetic patterns and it is the special cases that result from this function that predict the local precipitation of dolomite.

An interesting case arises when the slope of one boundary changes algebraic sign while the slope of the other boundary remains fixed (Figure 4). In a case such as this, the fluid following one boundary has a vertical component which changes sign with the boundary and this change in the vertical flow direction is reflected in the diagenesis.

Consider the following example. A pore fluid flow through a rock volume following an upper boundary which starts with a positive slope, changes to a negative slope and then changes back to a positive slope over some interval x. Over this same interval, the slope of the lower boundary remains positive (Figure 4).

---

[1]The only rational for this particular choice of stream function is that the flow follows the prescribed boundaries Z_b and Z_t and varies in a "rational" manner between these boundaries.

$$u = \frac{V_o}{Z_o - Z_t}$$

$$w = \frac{V_o}{Z_b - Z_t} \left[Z'_b + \left(\frac{Z_b - Z}{Z_b - Z_t} \right) (Z'_t - Z'_b) \right]$$

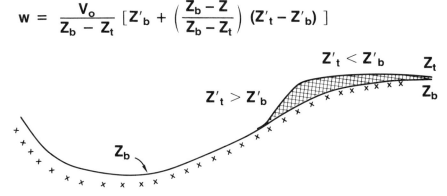

Figure 4 - Schematic structural configuration for text example. A sign reversal which occurs in vertical flow field is responsible for the development of the isolated pod of dolomite in the center of the carbonate matrix.

It follows from the previous discussion that there is a sub volume of that rock in which the pore fluid has a positive vertical component. This volume can be mapped out by setting $w = 0$ in Equation 16 and solving for z,

$$z = (Z_b - Z_t) Z_b' / (Z_t' - Z_b') + Z_b. \tag{17}$$

The volume of rock characterized by positive vertical flow will contain a different alteration assemblage than the rock characterized by a negative component of the vertical flow.

We can see this more easily in the case where the rock volume under consideration consists solely of quartz grains. In that case the volume of rock which sees the vertical flow component will contain precipitated quartz, predominantly as quartz overgrowths while the volume of rock which sees the negative flow component will be characterized by dissolution of quartz grains.

Diagenetic Fields

In the case of carbonate diagenesis, the distinctions are not as obvious. However, for the current model to be applicable, it will be necessary for the dolomite and calcite to have opposite behavior, in the sense that calcite will precipitate when dolomite dissolves and vice versa, as a fluid parcel crosses a temperature field. In particular we require that the thermal mass transfer coefficients (Equation 2) be of opposite sign, since this the only case in which one phase will dissolve as the other precipitates.

In the case of dolomitization, it is a further requirement that these coefficients be of equal, or approximately equal, magnitude,

$$\alpha_{T(cc)} = -\alpha_{T(dol)}. \qquad (18)$$

Under these conditions calcite will replace dolomite, or vice versa, with little or no volume change (i.e., isomorphic replacement), which is a typical observation.

For the purposes of this paper we will assume that the relation described by Equation 18 is true, at least approximately. We will not detail them here but calculations have been made which support this assumption.

TOPOGRAPHY

General Surface Function

The preceding discussion has shown that the geometry plays a significant role in diagenetic modification when pore fluid movement is involved. If the assumption is made that lithologic boundaries are effective flow boundaries then geologic structures become flow guides, channeling flow through strata in response to various potentials. In any real situation these boundaries are complex and vary in two dimensions. Consequently the simplest possible geometric model would consist of two surfaces, one which approximated the lithologic boundary between the granitic basement and the lower boundary of the carbonate layer and a second which approximates the contact between the upper carbonate boundary and the overlying shale or salt (Figure 1).

Mathematical models for surfaces of this type have not as yet received much attention in the literature, and consequently, it is not known what functions are most suitable. However, it would seem that any function which is differentiable and which permits easy location of domes and troughs would be suitable, particularly if complex topographies could be generated by simple addition of discrete functions.

It is suggested that a function of the form:

$$z(x,y) = \frac{S}{[(\frac{x}{a})^2 + (\frac{y}{b})^2 + 1)]} \qquad (19)$$

where S, a and b are arbitrary constants, is sufficient for present purposes. The general characteristics of this function are: (1) it has a maximum value of S at $x = y = 0$, and (2) it approaches a value of zero for large x or y (Figure 5). This function produces a circular dome for a = b and an elliptical dome for a <> b. It will produce a basin type of structure (anti-dome) for negative values of S.

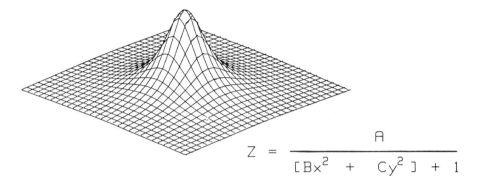

$$Z = \frac{A}{[Bx^2 + Cy^2] + 1}$$

Figure 5 - General surface function used to extend model to three dimensions. Z is given as a function of x and y according to the analytical expression shown. This surface is combined with linear combinations of other similar surfaces to approximate the boundary defined by the contact between the carbonate rock and either the upper shale/salt or the lower basement.

Mathematical Models For Surfaces

Upper surface

As an example we will develop a mathematical model for the upper and lower surfaces using Equation 19. Let

$$x = X - H_1 \qquad (20)$$

and

$$y = Y - K_1 \qquad (21)$$

denote translated coordinates, where H_1 and K_1 are constants and X and Y are the untranslated coordinates. Further let

$$S = -S_1$$

$$a = a_1$$

$$b = b_1.$$

Then we will define F_1 as:

$$F_1 = z(x,y) \ \{H_1, K_1, S_1^1, a_1, b_1\} \qquad (22)$$

where the contents of the curly braces, {P}, denote the parameters to be used in Equation 19 and P represents the set of parameters {H,K,S,a,b}.

Similarly the function F_2 is defined

$$F_2 = z(x,y) \quad \{H_2, K_2, S_2, a_2, c_2\} \tag{23}$$

and the lower boundary function, z_b, is defined as

$$z_b = F_1 + F_2. \tag{24}$$

Figure 6 is a representation of the lower surface used in the present model where suitable choices of the S, a, b, H and K parameters produce a topography representing a deep basin with an off-center ridge rising out of it.

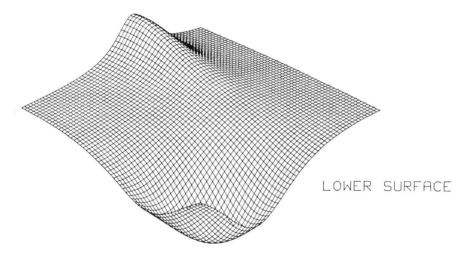

LOWER SURFACE

Figure 6 - A composite surface generated by adding two functions of the type shown in Figure 7 (one positive and one negative). This surface is used to represent the lower boundary between the carbonate and the basement. The "ridge" shown emerging from the top center of the surface represents the basement high depicted in Figure 1.

The values of the {P} parameters are somewhat arbitrary; any values which sufficiently separate the two elements are suitable. Alternatively, the surface can be modeled from a geologic or topographic map if desired. In addition, the individual elements, such a basin or a dome or ridge etc., can be rotated as well. Thus a rotation angle, α, can be added to the parameter list, P, which indicates the degrees of rotation of the axes of the element.

In this example the basin axis is rotated 30 degrees counterclockwise, relative to the ridge axis.

Upper surface
The upper surface of the model may be generated in a similar manner: we define the surface

$$z_t = F_1 + F_2 + F_3 \qquad (25)$$

where F_1 and F_2 are the functions defined above for the lower surface. F_3 is defined by

$$F_3 = z(x,y) \quad \{H_3, K_3, S_3, a_3, b_3\} \qquad (26)$$

and the {P} parameters are selected so that dome is added to the ridge (the F_2) in the lower surface (Figure 7).

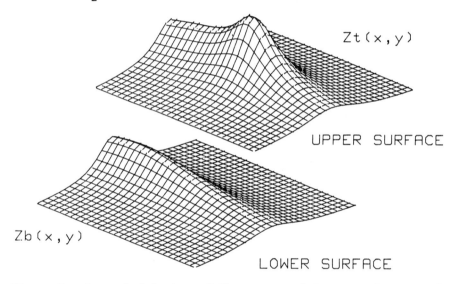

Figure 7 - An exploded view of the upper and lower surfaces used to model the dolomitization. The lower surface is a section of the surface shown in Figure 8 containing the "basement high" ridge, while the upper surface is the same as the lower surface with the exception of an additional term added to it to produce the dome or "hump" on the nose of the ridge. This perturbation in the geometry induces dolomitization in the vicinity of the dome crest.

Thus the only difference between the upper and lower surfaces is the extra function F_3 added to the upper surface. As will be shown below, this perturbation in the surface causes enough change in the flow field to cause precipitation of dolomite. The F_3 function perturbs the surface sufficiently to allow the fluid following the upper surface to flow upward as it ascends the dome and thus introduces a positive vertical component to the flow field locally. This region of positive flow defines the limits of the dolomitization, for the reasons discussed above discussed above.

Figures 8 and 9 are contour maps of the lower and upper surfaces respectively and clearly show the effect of the addition of the F_3 function. The closed contours on the nose of the ridge indicate a local high and we expect that the diagenesis in the vicinity of that high will be different for these two surfaces. In other regions, the two contour maps are essentially identical and we expect similar diagenetic patterns.

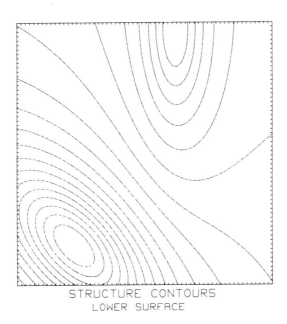

Figure 8 - Structural contours of the lower surface of the model. Compare with the diagenetic map of this surface (Figure 10).

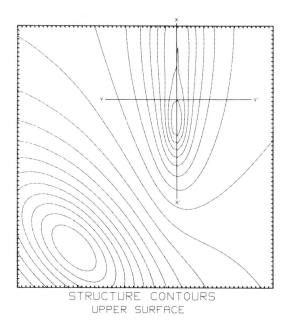

STRUCTURE CONTOURS
UPPER SURFACE

Figure 9 - Same as Figure 8 except that the structural contours are on the upper surface of the model. Compare with Figure 11.

DIAGENESIS

The diagenetic fields calculated for the two surface functions given by Equations 24 and 25 described above are shown in Figures 13 and 14.

Equation 16 gives the vertical component of the velocity field as a function of z for specified Z_b and Z_t. In order to contour the diagenesis on the <u>lower</u> surface we set $z = Z_b$ and compute

$$w = V_o Z'_b/(Z_b - Z_t) \qquad (27)$$

for given x and y on the map surface. We compute an array of w values and then contour them using a contouring algorithm. A similar procedure, setting $z = Z_t$, yields the diagenetic contours for the <u>upper</u> surface.

In order to calculate the diagenetic field, we must specify a direction of flow across the surface. We will obtain a different diagenetic field depending on the direction of flow. Intuitively, it makes sense that the spatial distribution of the diagenetic products should depend on the direction of the pore fluid movement and the directional derivative quantifies that notion.

Flow direction

For the purposes of this paper we will assume that the flow field is parallel to the long axis of the ridge in Figure 12 (parallel to the x-axis), and compute the diagenetic field relative to this flow direction. We will not explore the effect of varying the flow direction, but it should be obvious that the diagenetic field will shift in response to changes in direction.

Diagenetics Contours - Lower Surface

The diagenetic contours calculated from the directional derivative are plotted in Figure 10 for the lower surface. The blue contours indicate those parts of the surface which have a negative vertical flow component while the red contours indicate positive vertical flow components. That the diagenesis should behave in this manner can be verified by inspecting the surface in Figure 7 and noting that fluid flowing from top-to-bottom of will flow down hill in the blue region and up hill in the red region. Thus the vertical component of the flow field changes sign, and the predicted diagenesis responds according to Equation 13.

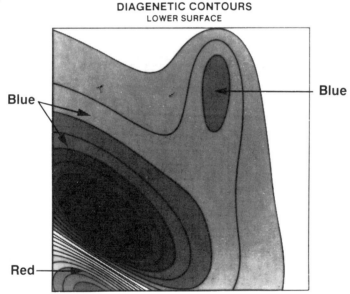

Figure 10 - Diagenetic contours on the lower surface of the model. The red contours indicate dolomite while the blue contours indicate calcite. The intensity of the colors indicates the intensity of the diagenetic alteration. Comparing this figure with the contour map of the lower contact shows clearly that the separation between calcite and dolomite is structurally controlled in this model. The alteration is least at the structural lows (and highs, see next figure) and tend to a maximum at the inflection points in the structural surfaces.

These two flow regimes distinguished by the sign change in the vertical velocity cause the diagenesis to be different in the two regions. The blue region will have a diagenetic signature that is different from the region covered with the red contours. On the lower surface the demarcation between the two regions coincides with one of the basin axes, the long axis in this case. The angle the fluid flow field makes with the long axis of the basin is evident.

The differing intensities of the colors in Figure 10 corresponds to intensity in the diagenetic alteration. For the lower surface, the heavy blue color indicate regions of relatively more intense calcite precipitation, while the deeper red colors indicate relatively higher degrees of dolomitization.

In these figures, the contours display the intensity of the diagenesis and represent constant values of w given by Equation 16. They thus represent values of the directional derivative divided by the thickness of the rock unit at one point and are not strictly proportional to fluid flux (rock-water ratios). The diagenesis is not, therefore, solely a response to the amount of fluid that has passed through the rock. However, because these contours are inversely proportional to thickness, they are proportional to rock-water ratio as well as the angle at which the fluid crosses an isotherm.

The diagenetic intensity is a result of two factors: (1) the angle the flow field makes with the temperature field and, (2) the relative fluid flux. A simple example is to note that one liter of fluid flowing parallel to the temperature gradient (at right angles to the isotherms) is equivalent to two liters of fluid flowing at 30 degrees to the gradient.

Diagenetic Contours - Upper Surface

The diagenetic contours for the upper surface are shown in Figure 11. The colors and contours mean the same things as in the previous figure for the lower surface. It is apparent that the upper surface has a diagenetic pattern strikingly different from the lower surface.

Both patterns are similar in the region of the basin but there are no similarities in the diagenetic patterns along the ridges.

On the lower surface the ridge is nowhere dolomitized while the red region in the upper surface indicates substantial dolomitization along parts of the crest and flanks of the ridge. The dolomitized area faces a down plunge area of intense calcite precipitation. Both of these regions contrast sharply with the diagenetic patterns in the lower surface and are a response to the presence of the small domed structure on the end of the ridge.

Cross Section Parallel to Flow

We can see the relationship between the structure and diagenesis more clearly by looking at a cross section through the ridge parallel to the long axis (Figure 12). The colors are consistent with the previous figures, red indicates dolomitization

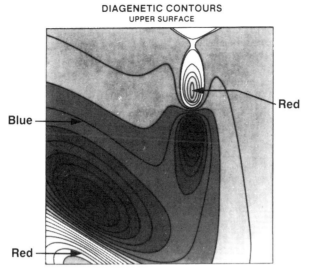

Figure 11 - Diagenetic contours on the upper surface of the model. The colors mean the same thing as on Figure 10. Note the alteration induced by the dome on the nose of the basement ridge, and how the dolomite is confined to a region slightly off the crest of the structure. Also note the presence of an intense region of calcite precipitation downstream of the dolomite.

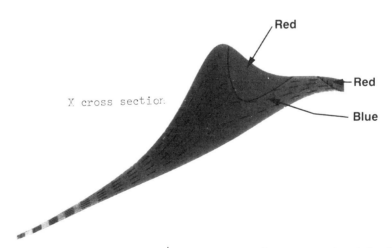

Figure 12 - Cross section XX^1 (see Figure 9) through the dolomite pod on the dome of the upper surface in the direction of fluid flow. The color codes mean the same as in the previous figures. The dashed lines trace the streamlines as the flow proceeds from right-to-left.

while blue indicates calcite precipitation. The boundary between the red and blue regions represents the locus of points for which w = o and was calculated using Equation 16 (the small black dots on the line represent calculated points). The dashed lines are streamlines and show the sense of fluid motion, out of the plane of the Figure toward the observer. Note that the upper and lower boundaries are streamlines and that the rest of the streamlines were calculated using Equation 14 in conjunction with Equations 24 and 25.

This cross section shows that the upper part of the unit is dolomitized while the lower part escaped dolomitization. The boundary between the red and blue is actually the locus of "zero" diagenesis and marks the location where the diagenetic character was changing over from alteration caused by a fluid moving upwards and cooling to alteration caused by fluid moving downward and heating.

Cross Section Perpendicular to Flow

The three-dimensional aspect of the altered rock can be appreciated by combining this cross section with Figure 13, which is a cross section at right angles to Figure 12. Note that the red dolomitized area appears to be two disconnected regions but this is because the dolomitization does not extend deeply into the rock volume along part of the boundary, and that this section is parallel to both the fluid flow and the long axis of the ridge.

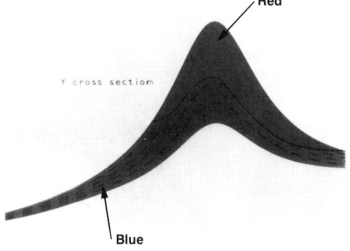

Figure 13 - Cross section YY^1 (see Figure 9) through the dolomite pod at right angles to the section shown in Figure 12. Colors mean the same as before and the dashed lines trace the flow as it comes out of the plane of the paper toward the observer.

The dashed lines represent streamlines with flow from right to left. Inspection of the streamlines in this cross section illustrates the point that the two regions are characterized by different vertical slopes on the streamlines.

The most intense dolomitization is evidently in the vicinity of the inflection point in the upper boundary, where the fluid flow turns upward most sharply. This is also true for the region characterized by intense calcite precipitation (dark blue contours in Figure 11).

SUMMARY & CONCLUSIONS

The principal conclusion resulting from this study has been to show that regional diagenesis can result from the slow movement of pore fluids and that the geometry, or structure, of the initial rock mass plays a critical role in the location and relative intensity of the alteration.

In the particular example considered here, the development of a pod of dolomite in a carbonate mass that was originally calcite, was caused by a slight change in the thickness of the carbonate unit. Because a small structural high was present over part of the rock mass, percolating pore fluids had to change direction over part of this region and this resulted in conditions suitable for dolomitization.

Such conditions are very common, and in fact, would be expected to be the rule over volumes of rock that cover regions as small as 10, or perhaps even 1, square kilometers. This model is thus applicable to virtually every large mass of carbonate that has undergone dolomitization in the subsurface. The basic requirements are:

1. The geometry of the rock body has sufficient irregularities to cause changes in the vertical direction of the fluid movement,

2. The fluid move slowly enough that chemical equilibrium can be maintained between the rock and the fluid, but fast enough to flux the equivalent of 10,000 to 100,000 pore volumes through the rock in the time available,

3. That the fluid has sufficient magnesium to saturate with dolomite, and that sufficient magnesium for the dolomitization be available.

Assuming that these conditions are met and are sufficient for dolomitization, we could reasonably expect that large masses of rock, on the scale of 10 to 100's of cubic kilometers, could be dolomitized to significant degrees over geologic time.

REFERENCES

Kaplan, W. and Lewis, D. J. 1970, <u>Calculus and Linear Algebra vol. 1</u>, John Wiley and Sons, New York.

Davis, S. H., Rosenglat, S., Wood, J. R., and Hewett, T. A. 1985, "Convective Fluid Flow and Diagenetic Patterns in Domed Sheets", Am. J. Sci., (285) p. 207-223.

Wood, J. R. and Hewett, T. A. 1982, "Fluid Convection and Mass Transfer in Porous Sandstones - A Theoretical Model", Geochim. et Cosmochim Acta 46(10) p. 1707-1713.

Wood, J. R. and Hewett, T. A. 1986, "Forced Fluid Flow and Diagenesis in Porous Reservoirs: Controls on the Saptial Distribution" <u>Roles of Organic Matter in Sediment Diagenesis</u>, D. L. Gautier (Ed.) SEPM Spec. Publ. No. 38, Tulsa, OK. U.S.A

Wood, J. R. and Hewett, T. A. 1984, "Reservoir Diagenesis and Convective Fluid Flow", in <u>Clastic Diagenesis</u>, D. A. McDonald and R. C. Surdam (Eds.) AAPG Memoir 37, Tulsa, OK. U.S.A.

T1 - PERMEABILITY CORRELATIONS

Roger Dashen, Peter Day, William Kenyon, Christian Straley, Jorge Willemsen
Schlumberger-Doll Research
Old Quarry Road
Ridgefield, CT 06877

ABSTRACT

We begin by presenting evidence which indicates a strong correlation exists between single fluid Darcy permeability and the T1 parameter characterizing the relaxation of nuclear magnetism within sandstones and carbonates. After defining terms, we discuss why NMR decay of water in rocks might reveal information about rock microgeometry. Finally, we discuss efforts being made to understand how to relate the length scale measured in NMR with that involved in permeability.

HIGHLIGHTS OF THE DATA

The principal purpose of this paper is to bring to the attention of this community some very interesting experimental results which have been obtained recently at SDR. Hopefully this will stimulate further research which can shed light on the questions raised by these results. We begin by providing a thumbnail sketch of the data which was generated, and on the methods used to analyze this data. Full details are available in Reference 1.

1) Porosity (ϕ), permeability (k), and T1 measurements were performed on a variety of sedimentary rock samples from different parts of the world. Figure 1 displays the correlation between these quantities that has been discovered[1].

2) The exponents of ϕ and of T1 were determined so as to minimize the errors in the fit. Earlier work[2] led one to expect that the exponent of T1 should be 2, while the exponent of ϕ was expected to be 1.

3) The figure shown is based upon a "stretched-exponential" representation of the data, i.e.,

$$M(t)/M_0 = e^{-(\frac{t}{T1})^a} \qquad (1.1)$$

It was found that this representation of the data had smaller fitting error for about 2/3 of the samples, while a two exponential fit ($n = 2$ below)

$$M(t)/M_0 = \sum_{i-1}^{n} M_i e^{-(\frac{t}{T1_i})} \qquad (1.2)$$

was better for about 1/3 of the samples.

By far the best fit among those attempted was for $n = 3$ in the formula above. However, while there may be some physical significance in the improved fit for some of the rocks, we are inclined to believe that in general the better fit merely reflects the greater flexibility for fitting which can be achieved with 6 parameters rather than 3 or 4.

In any case, we found that the overall goodness of the correlation is not sensitive to the choice of fit for T1.

4) Figure 2 shows the frequency of occurence of values of α. It is based only on those samples for which the stretched fit is superior to the two-component fit (although the other samples have α's in the same range). The shape of this histogram is suggestive, and we shall offer some ideas regarding its interpretation later.

We now pass to defining the quantities entering into the correlation more fully, especially k and T1, emphasizing which characteristics of the pore space of the rock these quantities most strongly depend on.

DARCY PERMEABILITY AND MICROGEOMETRY

The Darcy single phase permeability k is defined through the equation

$$q = \frac{k}{\mu} \frac{\Delta P}{L} \qquad (2.1)$$

In this equation, q is the volume flow rate per unit time averaged over the cross-section of the porous medium, μ is the viscosity of the flowing fluid, and $\frac{\Delta P}{L}$ is the macroscopic pressure drop across length L of the sample. The permeability is thus the "hydraulic conductivity" of the porous medium. It is not atypical for the permeability to be anisotropic, for example in rocks which are preferentially layered in one direction. For simplicity we shall speak of rocks as having "a" permeability, although in general the situation can be more complicated.

Examining the dimensions of the quantities in Eq.(2.1), one finds that the permeability k has dimensions of $(\text{length})^2$. For historic reasons, the oil industry uses the conventional unit

$$1 \text{ darcy} = 9.87 \times 10^{-9} cm^2 \approx 1 \text{ micron}^2 \qquad (2.2)$$

Sedimentary rocks such as sandstones and carbonates can have permeabilities anywhere in the range from micro-darcies to tens of darcies.

What properties of these materials can give rise to permeabilities in this range? A partial list of candidates includes sizes of pores and throats, interconnectedness of flow passageways through the material ("tortuosity"), chemical stability of the material (does any leaching occur over periods of flow, do clays swell, etc.?), and mechanical integrity of the substance (are small particles swept through which can clog the flow downstream?). Our discussion will be entirely limited to consideration of the effects of distributions of pore sizes on permeability.

Within this framework - permeability depends solely on pore sizes - sedimentary porous materials can be described in terms of a network model of the pore space, which consists of "bonds" representing fluid passageways across the lengths of grains, and "sites" representing intersections of such passageways, roughly corresponding to corners where several grains meet.

The permeability of the entire sample can then be calculated making use of this network model. One assigns a site-to-site permeability along each bond with a value corresponding to a size chosen so that the passageway behaves like a tube of that size. It is reasonable to focus on the tubes rather than on the pores simply because it is hard to flow through small constrictions. The sizes are assigned according to some probability distribution, which might be chosen on the basis of independent measurements, or on the basis of computational convenience.

Having done this, one solves for the effective permeability of the network, given the permeability of its components[3]. This is a standard problem in statistical physics, and there are many approximation methods available to tackle it[4,5].

The key point is that these statistical methods enable one to quantify the permeability in terms of microgeometrical length scales associated with the statistical distribution of pore and grain sizes.

WHAT IS T1?

For our purposes here we will need to touch upon only the most rudimentary facts regarding NMR[6]. The first fact is that the protons which are in the hydrogen which is in the molecules which comprise water are endowed with nuclear magnetic moments. These moments enable them to preferentially align with an externally applied magnetic field, like so many iron filings.

But the molecules of water at room temperature are in constant thermal agitation, as though the iron filings were constantly being shaken about. This means first that not all of the protons are aligned with the external field, only a few of them, yet fortunately enough to be observable. Thermal agitation also means that once the external field is removed, the system will return to its earlier equilibrium configuration, in which there is no net magnetization of the protons at all.

It turns out that the time it takes for the system to return to thermal equilibrium after removal of the field is also the time it takes for the system to build up to its maximal magnetization starting immediately after the external field has been established. This "build-up" time is conventionally denoted "T1", and is called the "longitudinal" relaxation time, referrring as it does to the time required for the nuclear magnets to "relax" either into alignment along the external field, or back to random alignment in the absence of that field.

In practice, there exists a hefty arsenal of experimental techniques which have been introduced to measure times characterizing nuclear magnetic relaxation. T1 is only one such time, but it is the only one we will need for today.

The central question to be addressed is how this time can be related to the (distance)^2 which characterizes the permeability.

The answer to this hinges on one further important fact: the T1 measured for water saturating a porous material is observed to be much smaller than it is for that same water in bulk. The standard interpretation of this reduction in T1, which can be demonstrated experimentally, is that T1 for water is affected significantly by surface relaxation mechanisms. It is in this manner that the decay constant can be related to the geometrical structure of the porous material. We discuss this relationship in detail in the next section.

MODELLING RELAXATION IN ROCKS

In this section we begin by stating well-known results regarding the effect of surface decay on T1 for simple geometrical objects such as cylinders and spheres. We then try to extend the intuition gained from these studies to the intractable problem of solving the relevant equations within the full pore space of a rock. A reasonable-seeming scenario emerges, but it remains one which needs to be checked through extensive modelling.

A. Surface decay in isolated containers

The basic phenomenon to be described is this[7]: water molecules are aligned with an externally applied magnetic field and it is assumed that at $t = 0$ a spatially uniform magnetization has been achieved. For times greater than 0, the field is turned off, and magnetization decays according to the bulk rate, but in addition due to interactions with the walls of the container. Encounters with the wall occur in the normal course of molecular diffusion, so we write

$$\frac{\partial M}{\partial t} = D\nabla^2 M - \frac{1}{T1_b}M \qquad (3.1)$$

where D is the diffusion constant for water, and $T1_b$ is the bulk decay constant for water. We can solve for the bulk contribution to the time dependence immediately, and we will eliminate this term from our discussion entirely in what follows.

The decay at the wall is entirely described by the phenomenological parameter ρ:

$$D\nabla_n M + \rho M)_{wall} = 0 \qquad (3.2)$$

Notice that parameter ρ has dimensions of length/time, corresponding physically to an interaction range times a relaxation rate. Microscopic theory should describe this parameter, but for purposes of describing the macroscopic decay, ρ is simply an adjustable parameter of unknown origin[8].

The equations above may be solved in closed form through conventional methods for simple geometries such as parallel plates, cylinders, and spheres. One obtains generically

$$M(t) = \int dV\, M(\vec{x}, t) = M_0 \Sigma_n I_n e^{-\lambda_n t} \qquad (3.3)$$

Here the amplitudes I_n are chosen to satisfy the initial condition of uniform magnetization ($\Sigma_n I_n = 1$), and the decay rates λ_n are obtained by solving transcendental equations stemming from the boundary condition. It is useful to write the rates in the form

$$\lambda_n = \frac{D}{a^2} f_n\left(\frac{\rho a}{D}\right) \qquad (3.4)$$

For reasons which will be evident later, we will call function "f" the "transformer" function. Inspection of the functions f_n for the simple geometries cited reveals the following general scenario:

1) Surface limited decay

If the dimensionless ratio $\rho \frac{a}{D} \ll 1$, a single mode is sufficient to describe the decay, $I_0 \approx 1$, and f_0 is such that $\lambda_0 \approx \frac{\rho S}{V}$, where S/V is the surface to volume ratio of the container.

2) Diffusion limited decay

In the opposite limit $\frac{\rho a}{D} \gg 1$, the wall kills off the magnetization just as rapidly as the bulk can supply it through diffusion, and thus the diffusion process limits the rate of decay: $\lambda_0 \approx \frac{D}{a^2}$. Unlike the first case, however, I_0 is less than 1, reflecting that at early-times the magnetization that was initially in the vicinity of the wall was depleted rapidly. Indeed the contributions of the higher modes can naturally be interpreted as transients which describe the change in spatial distribution from uniform at $t = 0$ to one in which diffusion supplies particles to the wall at just the rate at which the wall can demagnetize them.

3) Intermediate case

When the two limits above are not realized, that is, $\frac{\rho a}{D}$ of order 1, the asymptotic rates quoted above remain reasonable estimates of the dominant rate. The main difference is that the transient behaviour lasts relatively longer than in the extreme cases, and so a multi-exponential description remains appropriate throughout most of the observable decay.

B. General remarks on the rock response

Passing now to the complicated geometry of the pore space, we assume that equations (3.1) and (3.2) still apply. In particular, the diffusion constant D is that for bulk water, and a single parameter ρ applies uniformly throughout the rock to account for surface relaxation. The first assumption is based on the observation that T1's in rocks are of the order of 100's of milliseconds. Assuming $D = 2 \times 10^{-5}$ cm.^{2} / sec., a molecule diffuses appx. 35 microns in that time. It does not explore a lot of the rock, and consequently tortuosity corrections to the bulk water diffusion coefficient are not needed. The second assumption is just the simplest thing to do in the absence of contrary information.

While we do not know how to apply the boundary condition (3.2) for the full material, we expect nonetheless that the final answer for the magnetization will have the structure of Eq.(3.3). The rates λ_n will depend upon D, ρ, and all the details of the porespace microgeometry. Let us proceed in a standard way, and replace the summation over modes in (3.3) by an integral, thence to an integral over relaxation rates:

$$\frac{M(t)}{M_0} = \int d\lambda P(\lambda) e^{-\lambda t} = \int d\tau P(\tau) e^{-\frac{t}{\tau}} \qquad (3.5)$$

The function $P(\lambda)$ is thus a density of modes weighted by the amplitudes needed to satisfy the initial condition. The second form merely involves a change of variables from rates to time constants.

At this point recall that "stretched" exponential decay is often a good representation of the data[1]. Plausibility arguments for the success of this representation can be given based upon Eq. (3.5). As a simple example, imagine momentarily that the decay time constants τ in the porous material arise from every conceivable random process: random pore sizes, random shapes, random ρ values, random anything else you can think of. It might then happen that the distribution of decay constants follows the normal distribution, by the central limit theorem:

$$P(\tau) = \frac{2}{\sigma\sqrt{\pi}} \frac{1}{1 + erf(\tau_0/\sigma)} e^{-((\tau-\tau_0)/\sigma)^2} \tag{3.6}$$

In this case, a steepest descents estimate of the integral in Eq.(3.5) yields the result

$$M(t) \rightarrow e^{-(3^{3/2}t/2\sigma)^{2/3}} \tag{3.7}$$

in the limit that $t \gg \tau_0^3/\sigma^2$.

We cannot be certain that it is the τ's that get normally distributed and not the λ's. Nonetheless this simple model puts us into the range of values for α shown in Fig.2.

The example above can be generalized, an exercise which leads to the following observations:

1) The stretch exponent α is related to the rate at which the distribution of lifetimes falls off for large τ:

$$P(\tau) \approx e^{-(\tau/\tau_0)^{\alpha/1-\alpha}}$$

2) The observed lifetime $T1_{\text{exp}}$ is proportional to τ_0 and is thus intimately related to the width of the distribution.

That is, stretched behaviour can emerge in a natural way from the existence of a broad distribution of relaxation times peaked toward small τ. Given that for simple geometries the representation Eq. (3.3) encompasses transient modes, and for the full rock we must fold in random sizes and shapes, it would appear most *un*-natural for $M(t)$ to be the sum of a small number of single exponential terms. Nonetheless sometimes that may happen, too, and below we shall indicate conditions under which it might be so.

It is possible to go a step further and work out a procedure for "inverting" Eq.(3.5) to obtain $P(\tau)$ from the data when stretched exponential is a good fit[9]. We cannot describe the details of this approximate inversion scheme here, but Figure 3 provides an illustration of the method applied to a real rock sample. A virtue of the inversion procedure over general methods which might be tried when stretched-exponential is not a good fit to the data is that it is demonstrably stable to first order in changes in the data, e.g., noise fluctuations.

To summarize this subsection, we can pose the relaxation problem for the porous material in a well-defined manner, but we cannot solve it analytically as one can for simple geometries. Nevertheless, the stretched decay observed in a substantial number of rocks can be readily interpreted in terms of a broad distribution of decay rates, which one expects in a complicated material. The difficult problem which remains is to identify the spread of this broad distribution with a characteristic length.

C. How can T1 tell us about sizes?

It would appear that by one more change of variables, from τ to "a", we can extract an NMR determined "pore-size distribution". But how do we make this change of variables? For the simple geometries, the required transformation was effected through Eq.(3.4). Recall that the precise form of the transformation function f depended upon the specifics of the geometry, but in all cases it had the same general features. So why not postulate a transformer with the same features and proceed with that as a further assumption?

We state two further results from Ref. 9 which bear on this question, and then comment on their relevance:

1) As $\rho \to 0$,

$$\lambda_0 \to \frac{\rho S}{V} + \text{order}(\rho^2) \tag{3.8}$$

Here λ_0 is the longest lived mode, S is the surface area of the medium, V its volume.

2) As $\rho \to \infty$,

$$\lambda_0 \to \lambda_\infty \left[1 - \frac{D}{\rho} \frac{\int dS (\partial M_\infty / \partial n)^2}{\int dV (\nabla M_\infty)^2} + \text{order}(\rho^{-2}) \right] \tag{3.9}$$

where λ_∞ is a constant which, like M_∞, is obtained by solving Eqs.(3.1,3.2) with D set equal to zero in the boundary condition.

These results indicate the same trends as the asymptotic expressions for simple geometries. If ρ is very small, we expect to see a long time behaviour entirely dominated by a single exponential decay, with the rate proportional to the surface to volume ratio.

If ρ is very large we likewise expect a single long-time mode, but unfortunately we do not know λ_∞. Nonetheless, it is instructive to examine the structure of the first correction term in Eq.(3.9). We argued earlier that in the diffusion limited regime the long-time spatial configuration of the magnetization contains gradients from the center of the pore to the wall. These have just the right magnitude that diffusion supplies victims to the wall at the rate at which the wall consumes them. Thus in some crude way we might write

$$\int dS (\partial M_\infty / \partial n)^2 = \Sigma_{\text{pores}} (\text{surface area}) \times \left(\frac{1}{\text{pore width}} \right)^2$$

Similarly we might consider the volume integral in Eq.(3.9) to consist of two pieces, one directly analogous to the above surface integral, a second including genuine volumetric gradients only. How would these second gradients originate? If a small pore is next to a big pore, we might expect the smaller one to decay faster than the larger, thus creating a gradient from pore to pore which is not restricted to the wall. Figure 5 illustrates this effect from a computer simulation. Then in the same crude way,

$$\int dV (\nabla M_\infty)^2 = \Sigma_{\text{pores}}(\text{surface area}) \times \left(\frac{1}{\text{pore width}}\right)$$
$$+\Sigma_{\text{pores}}(\text{volume}) \times \left(\frac{1}{\text{pore length}}\right)^2$$

These equations are based on the assumption that "unnecessary" gradients have been smoothed out and the remaining gradients reflect the microgeometry of the medium.

We learn two things from this caricature of the correction term in Eq.(3.9). First, the bad news: it is evident that the long term behaviour indeed involves averaging over the pore space size distribution, but notice that the averaging occurs in defining the rate λ, not in averaging decay terms over distinct rates λ, one for each pore size.

On the other hand, the computer simulation which purportedly clarifies the origin of volume gradients also suggests that at long enough times, "diffusional isolation" sets in, with smaller pores having died away, and with small chance of a diffusing bit of magnetization to make its way from a large pore to another large pore without being destroyed in the intervening small throat.

To the extent that the above suggestive physical picture is valid, with the magnetization having settled into diffusionally isolated pockets, it may be reasonable to associate individual time constants τ with individual pore sizes "a" provided we have settled past the high-mode transients. We can revert to the simple form Eq.(3.4) (with $n = 0$) since it is certainly dimensionally correct provided we keep in mind that any transformation from times to sizes involves guessing the transformer, f. It can be different from the form derived from simple geometries because the long-time spatial configurations are much more complicated.

In what follows we shall adhere to this optimistic outlook, although this is certainly one area where further investigation is required. It is entirely possible that the structure of the correction term in Eq.(3.9) is the tip of an iceberg which threatens to rip apart our usual way of modelling averages over the porous medium.

THE COMPONENTS OF A PERMEABILITY-T1 RELATION

Our aim now is to understand what physics might underlie the equation

$$k \approx \kappa \phi^p T1^q \qquad (4.1)$$

We first write all quantities in this expression in terms of lengths which relate to the microgeometry of the pore space. It will turn out that no unique length scale emerges from this procedure. Rather, we will find that Eq.(4.1) contains transport lengths, lengths characterizing the porosity, grain sizes, and the length scale determined by NMR.

The second step in the argument will invoke relationships among some of these different lengths which have been established through study of electrical conductivity through porous media[10,11].

The biggest problem in providing a basis for Eq.(4.1), however, lies in determining the form of the transformer function f, which relates NMR times to length scales. Some of the ideas introduced in the previous section will be brought to bear on this problem, and a new approach to parametrizing the transformer will be reviewed briefly[12].

Finally, some speculative ideas are introduced which might in principle lead to a determination of an effective transformer, based on self-consistency of the approximations leading from times to sizes.

A. Express all quantities in terms of microgeometrical lengths

To start, working within a standard statistical model of the type mentioned earlier, write the following general expression for the permeability of a porous medium:

$$k = k_0 \frac{R_4}{l^2} \qquad (4.2)$$

Here "l" is to be thought of as a typical pore-to-pore channel length, or as a typical grain size. R_4 refers to a quantity with dimensions of length to the fourth power, derived from the distribution of pore and throat sizes put into the model of the porous medium. The choice of method for calculating R_4 is not relevant to our current purpose.

Next, write the T1 obtained from a rock in the form of Eq.(3.4),

$$T1 = \frac{a_N^2}{D} f^{-1}\left(\frac{\rho a_N}{D}\right) \qquad (4.3)$$

Here a_N represents a parameter with dimensions of a length which characterizes the rock T1 decay.

Consider now the porosity ϕ, defined as the ratio of pore volume to the sum of pore and grain volumes. To work with a concrete example of the notions to be presented, consider a network of tubes consisting of radii drawn from some distribution, and of length l (same l as in Eq.(4.2)), and write symbolically

$$\phi \approx \frac{a_p^2 l}{a_p^2 l + l^3} \qquad (4.4)$$

Here a_p represents a pore radius within the distribution which best characterizes the distribution when calculating the porosity. To make our bookkeeping even simpler, let us further assume that $l \gg a_p$ so that

$$\phi \approx \frac{a_p^2}{l^2} \qquad (4.5)$$

One is not expected to believe a word of this in detail for a rock, the idea is to manipulate simple symbols which help us to remember the qualitative features of the quantities involved.

We are now ready to put the pieces together and replace Eq.(4.1) with the expression

$$k_0 R_4 \approx \kappa a_p^{2p} a_n^{2q} \frac{f^{-q}\left(\frac{\rho a_N}{D}\right)}{l^{2(1-p)} D^q} \qquad (4.6)$$

The first thing to notice about this equation is that the left hand side depends entirely upon the microgeometry of the medium while the right hand side depends upon that and in addition on totally "external" parameters ρ and D. This will be our central concern.

However, there is a second peculiar aspect to the equation. Notice that the microgeometric parameters appearing on left and right sides, R_4, a_p, a_N, and l, are in principle determined from different features of the probability distribution of sizes. An equation such as this can be satisfied only if the probability distribution has a rather special form[10,11]. Let us probe this in greater detail.

B. The conspiracy of lengths

Consider Archie's Law for the electrical conductivity of a rock saturated with saline:

$$\frac{\sigma_{\text{rock}}}{\sigma_{\text{water}}} = \phi^m \tag{4.7}$$

Statistical arguments like those leading to Eq.(4.2) give a result

$$\frac{\sigma_{\text{rock}}}{\sigma_{\text{water}}} = \frac{R_2}{l^2} \tag{4.8}$$

where R_2 is the same type of statistical microgeometrical quantity as R_4 of the permeability, but having dimensions of length squared.

Equating right hand sides of the preceding 2 equations and using Eq.(4.5) we have

$$R_2 = a_p^{2m} l^{2(1-m)} \tag{4.9}$$

From the mathematical point of view, this implies certain relationships among properties of the probability distribution of pore sizes. To the extent that the Archie exponent "m" lies within a narrow range for a broad class of rocks, these rocks share these relationships.

All of which leads us to the conclusion geologists reached long ago, namely that rock microgeometry is not purely random, but has some elements of commonality, presumably due to the processes through which the rocks were formed.

We see, then, that a "conspiracy of lengths", necessary for the desired correlation of permeability with T1 to work, is not new. One such appears also in Archie's law for the electrical conductivity, and it is assumed to originate from the diagenetic processes the rock has undergone. Which is not to say, however, that these processes are understood.

C. The external parameters

Returning to our examination of Eq.(4.6) let us agree for the purposes of argument that the required relationships among the microgeometrical parameters can be satisfied because the rocks for which the equation holds share certain unspecified structural characteristics. But unlike the case of Archie's law, we still have to get rid of the unpleasant ρ and D factors in the equation. Indeed, another way of phrasing things is that we must understand why κ, like Archie exponent m, does not vary much from rock to rock.

Based upon our discussion of T1, we must consider at least the following possibilities in order to unravel this independence:

1) Parameter ρ is very large in the rock samples

As mentioned earlier, this implies that $f(\rho a_N/D)$ approaches a constant value, independent of ρ. The surface decay is then limited by molecular diffusion, ρ effectively drops out of the picture, and $\kappa \approx D^2$ implies that κ is constant from rock to rock because D for water doesn't vary in the experiments.

Returning to Eq.(4.3), this also implies that T1 is ρ independent. However, there is evidence that T1 does depend on ρ. This can be checked separately because ρ depends upon microscopic relaxation rates at which the surface mechanisms operate. In general one expects such microscopic rates to depend upon the field strength at which the experiment is conducted, for reasons which we will not dwell on here. If the field strength is varied, T1 is observed to change. Thus T1 cannot be ρ independent, and this option is ruled out experimentally[13].

2) Parameter ρ is very small

In this situation $f \approx \frac{D}{\rho a_N}$. This eliminates D from the problem, and we are left with $\kappa \approx \rho^2$. If κ does not vary much from rock to rock, ρ cannot vary much from rock to rock.

While we have no evidence to rule this out, neither do we have independent evidence to confirm it. Thus this possibility must remain open, awaiting both theoretical and experimental testing. For instance, preliminary computer simulations suggested by one of us (W.K.) indicate that surface roughness may play an important role in weakening the apparent strength of the surface decay parameter. That is, one starts a calculation inserting a value for the parameter ρ, and then finds that the decay behaves as though a different, "effective" ρ, was in operation provided the surface is sufficiently rough. Further details of this investigation will be published separately.

3) The transformer function f isn't what we think it is

A rather clever idea has been put forth recently[12] which is too lengthy to relate in full here, but which in cartoon form says the following:

Suppose $f^{-1} = \left(\frac{\rho a}{D}\right)^x$. When $x = 0$ we are diffusion limited, when $x = -1$ we are ρ limited. Suppose further[10,11] that the statistics R_4 and R_2 are completely dominated by a single size a_0. Archie's law in the form Eq.(4.9) then implies that $a_0 = a_p{}^m l^{(1-m)}$. (As an aside, one can then write $k \approx \phi^{2m} l^2$.)

Making use of these relations, we can rewrite Eq.(4.6) in the form

$$\frac{\kappa}{k_0} \approx \frac{a_p^{2(2m-p)} l^{2(p+1-2m)} D^{q(1+x)}}{a_N{}^{q(x+2)} \rho^{qx}}$$

One can then "optimize" the formula by requiring that two of the unkown lengths, a_N and a_p, cancel. Forcing these lengths out of the formula means the only length scale left is the "grain size" l, thereby reducing our ignorance. NMR is still in the formula because ρ and D are. Inserting experimental numbers,

$$\frac{\kappa}{k_0} = l^{(10.6-4m)} D^{2.3(1+x)} \rho^{2.3x}$$

If x is close to but not equal to 0, κ is relatively insensitive to ρ, yet T1 retains a ρ dependence.

Unfortunately, however, the above optimization requires that exponent $x = \frac{2}{q}(2m - p - q)$. The proposed range of x imposes the restriction that $p + q >$

$2m > p+\frac{q}{2}$, which is not realistic for rocks, which have Archie exponents roughly between 1.5 and 2.5.

Putting aside this detail of the model, the positive point is the suggestion that we should treat the transformer function f as a "free parameter", so-to-speak, adjusting its functional form through scaling arguments to obtain the desired correlation.

D. Can we compute the transformer indirectly?

We now make use of the above suggestion in a speculative way. In detail, the transformation of the full magnetization from an integral over lifetimes to an integral over sizes involves the Jacobian factor

$$d\tau P(\tau) = da \frac{d\tau}{da} P(a(\tau)) \tag{4.10}$$

The arguments of the preceding sections suggest that the $P(a)$ constructed in this fashion can at best reflect the "true" distribution of pore sizes only for large sizes, corresponding to times when the pore space has become diffusionally compartmentalized. Assuming optimistically that indeed $P(a(\tau)) \to P_{\text{geometrical}}(a)$, we can demand

$$\frac{\partial P(a(\tau))}{\partial \rho} = \frac{\partial \tau}{\partial \rho} \frac{\partial}{\partial \tau} \left[\frac{\partial a(\tau)}{\partial \tau} P(\tau) \right] = 0 \tag{4.11}$$

$$\frac{\partial P(a(\tau))}{\partial D} = \frac{\partial \tau}{\partial D} \frac{\partial}{\partial \tau} \left[\frac{\partial a(\tau)}{\partial \tau} P(\tau) \right] = 0 \tag{4.12}$$

Without entering into details, it is clear that this will lead to nonlinear second order differential equations for f. It will be difficult to solve these equations *because* they are nonlinear, second order, simultaneously required, and must lead to f positive definite.

Further, why should f depend upon $P(\tau)$? To the extent that the physical picture regarding formation of pockets of magnetization is valid, this is reasonable because some rocks will achieve the desired configuration more quickly than others. If the contrast between pores and throats is large and the number of throats per pore is small, diffusional isolation may set in more quickly than in the opposite case. Such differences will show up in $P(\tau)$. They show up in f because a priori we don't know how far along in the process of diffusional isolation we really are, and by imposing Eqns.(4.11) and (4.12) we are hoping to find the best possible transformation from times to microgeometry.

SUMMARY AND CONCLUSIONS

Our starting point in this paper was the observation that there exists a good correlation between Darcy permeability k and the function $(\phi^4 T1^2)^{1.13}$ for a large number of sandstones and carbonates.

Although the correlation is relatively insensitive to the manner in which T1 is extracted from the data, it was found that for many of the samples a stretched exponential representation of the decay of magnetization is better than a 2-component representation. This form can arise when the distribution of decay times within the rock is peaked at very small times, but is itself stretched out to very long times. Such a scenario seems to us plausible for materials with as much heterogeneity over the scale of pore and throat sizes as rocks have.

In seeking to understand the observed correlation, we noted that each parameter in the correlation depends upon distinct microgeometrical length scales. Some of these are interrelated through relationships attributable to diagenetic processes. However, there remains a fundamental problem in relating the NMR length a_N to the other lengths in the problem in a manner which is consistent with the observed correlation and with other data[13,14].

The fundamental problem noted above stems from the inherent difficulty in solving the diffusion-decay equations (3.1) and (3.2) within the pore space of a rock. While simple geometrical models are useful in supplying insight into the physical processes at work, the transformation of decay time distributions to pore size distributions requires physical modelling which has not been fully established.

We noted, for example, that asymptotic expansions for the decay rate (Eq.(3.9)) suggest that averaging the rate λ over the pore space may be more correct than averaging $e^{-\lambda t}$, a situation which has arisen in other physical contexts. On the other hand, we noted that the physical picture of diffusionally isolated pocket formation might be appropriate for certain rock geometries. This picture would support assuming there exists a one-to-one correspondence between a rate λ and a size. Yet even this picture may be significantly affected by the existence of surface roughness, with possible renormalizing effects on the wall strength ρ.

Faced with the above uncertainties, we have proposed that one seek a "best" transformer function appropriate for a given set of rocks, rather than proceeding on the assumption that there exists a unique such function. The scaling ansatz put forth in Ref. 12 is one example of such a search. We have supplied a different, although cumbersome, method in the preceding subsection, which is based on self-consistency and reflects directly the assumptions which go into constructing such a transform.

In conclusion, it is amply evident that our current state of understanding is not yet at a level which permits us to "explain" the experimental results. If anything, we hope that we have helped sharpen the questions which need to be answered before substantial improvement in our predictive power can take place. This hope extends to sharpening the analysis of the data itself, to characterizing the rock samples by more than k, ϕ, and T1.

ACKNOWLEDGEMENTS

We thank Jayanth Banavar and Larry Schwartz for continuous stimulating interaction, and most especially Max Lipsicas for his warmth, his humor, his insight, and his sharp questions.

REFERENCES

1. W.E. Kenyon, P.I. Day, C. Straley, and J.F. Willemsen, SPE paper 15643, presented at 61st Annual Technical Conference and Exhibition of the Society of Petroleum Engineers, Oct. 5-8, 1986.

2. A partial list of early contributions includes D.O. Seevers, SPWLA Trans. (1966) Paper L; A. Timur, Trans. 9th Annual SPWLA Symposium (June 1968); A. Timur, The Log Analyst, (Jan-Feb 1969) 3-11; A. Timur, JPT (1969) 775-786; J.D. Loren and J.D. Robinson, Soc. Pet. Eng. J. (1970) 268-278; J.D. Loren, J. Pet. Tech. (Aug. 1972) 923-928.

3. J. Koplik, J. Fluid Mech. 119, 219 (1982).

4. An alternative to the effective medium theory approach used by Koplik is discussed in V. Ambegoakar, B.I. Halperin, and J.S. Langer, Phys. Rev. B4, 2612 (1971).

5. Both of the above methods are reviewed by S. Kirkpatrick, "Ill-Condensed Matter", ed. R. Balian R. Maynard, and G. Toulouse, North-Holland (N.Y.), 1979.

6. A good introduction to an extensive literature on studying porous media using NMR techniques is K.S. Mendelson, J. Electrochem. Soc. 133, 631 (1986).

7. S.D. Senturia and J.D. Robinson, Trans. SPE, 249, 237 (1970); K.R. Brownstein and C.E. Tarr, Phys. Rev. A 19, 2446 (1979).

8. When T_2 is in question rather than T_1, a method has been proposed to measure the corresponding ρ experimentally through the use of pulsed gradient methods. M. Lipsicas, J.R. Banavar, and J.F. Willemsen, Appl. Phys. Lett. 48, 1544 (1986).

9. R. Dashen and J.F. Willemsen, in preparation.

10. P.-Z. Wong, J. Koplik, and J.P. Tomanic, Phys. Rev. B 30, 6606, (1984).

11. J.N. Roberts and L.M. Schwartz, Phys. Rev. B 31, 5990 (1985).

12. J.R. Banavar and L.M. Schwartz, SDR preprint, Sept., 1986.

13. M. Lipsicas, private communication.

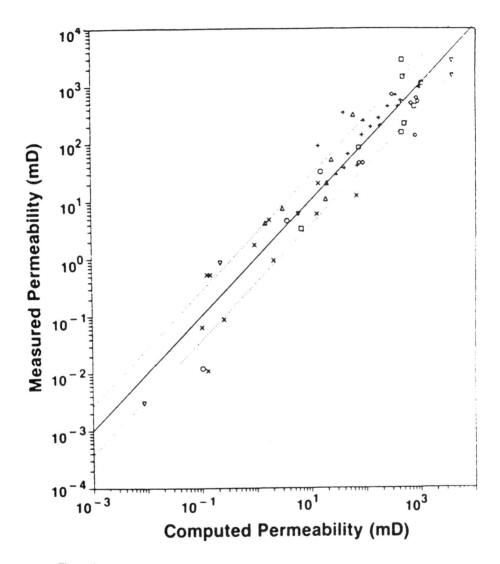

Figure 1
Measured permeability vs. the fit $k = (1.6 \times 10^{-9}) T_{1,\alpha}^{2.3} \phi^{4.3}$. This figure does not include two sample sets which fall well outside of the dashed lines. Closer examination has shown these sets to be anomalous in other ways as well.

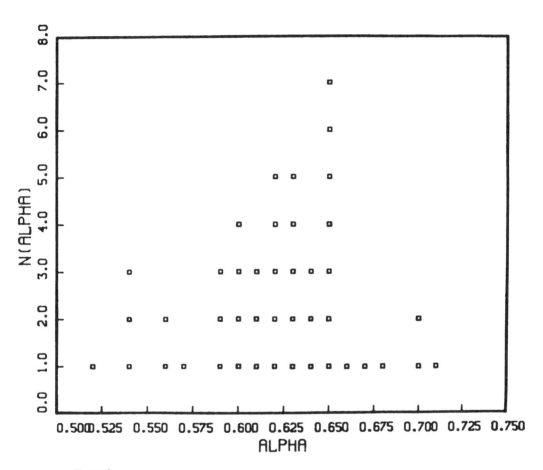

Figure 2
The frequency distribution of stretch exponents α for those samples for which the stretch fit is better than a 2-component fit. While statistics are clearly poor with only 43 data points in all, there is a trend to the distribution.

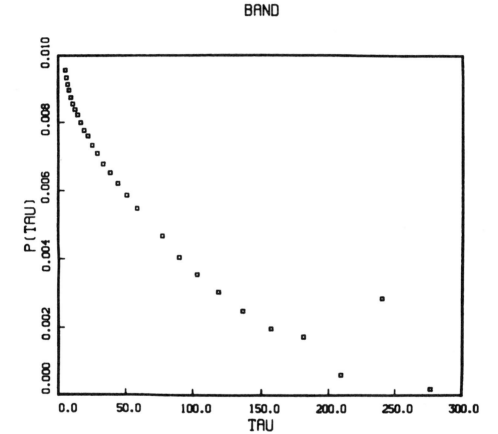

Figure 3

An example of a lifetime distribution estimated from the data. Here τ is measured in milliseconds. The accompanying Fig. 4 shows the goodness of fit to a stretched exponential for this particular sample. (The scatter at large τ comes about because the inversion as illustrated is based on the actual data). The inversion procedure used to produce $P(\tau)$ can be re-inverted to reconstruct the data. Estimates on the validity of the derived $P(\tau)$ can then be made based upon comparison of the reconstructed data with the original data.

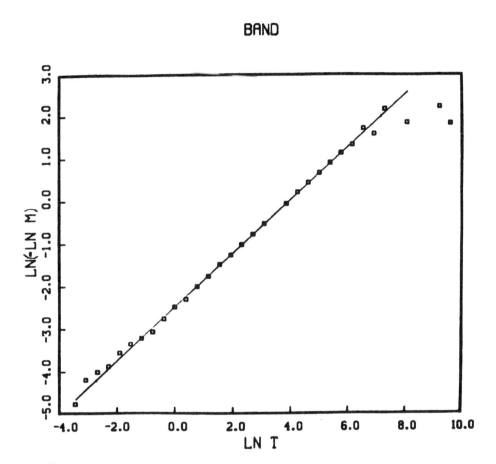

Figure 4

Decay data for a Bandera sandstone. This format best illustrates the validity of a stretched exponential fit (solid line). A slope of 1 would indicate single exponential decay. (Times are in milliseconds, and ln is natural log.)

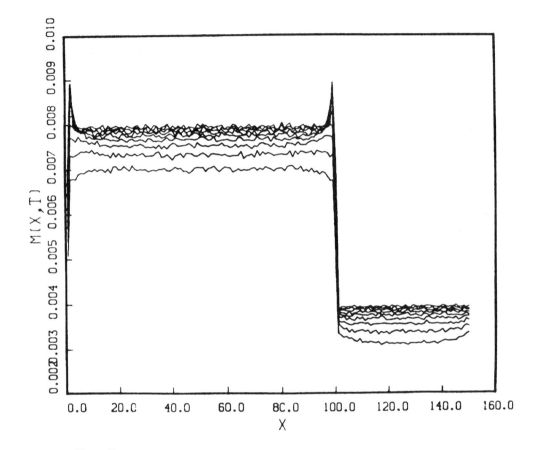

Figure 5

Preliminary results on the magnetization profile in a two-box system. Normalization of $M(x,t)$ is arbitrary. This profile was generated by simulating random walks and surface decays in two squares, one twice the length of the other. Boundary conditions at far left and far right are periodic, and surface decay occurs only at the top of the boxes, other boundaries being reflecting. $M(x,t)$ is the integral across the box from top to bottom. The sequence of curves is smallest times at the top, largest at the bottom.

ORIGIN OF POROSITY IN SYNTHETIC MATERIALS*

D.W. Schaefer, J.P. Wilcoxon, K.D. Keefer, and B.C. Bunker
Sandia National Laboratories
Albuquerque, NM 87185

R.K. Pearson, I.M. Thomas, and D.E. Miller
Lawrence Livermore Laboratory
Livermore, CA 94550

ABSTRACT

Several porous silicas have been studied in order to determine the origin of porosity in random porous media. The silica system offers a unique opportunity to study the origin of porous structures because a variety of different structures can be synthesized depending on precursor chemistry and physics. In solution-grown materials, for example, it is possible to grow particles with randomly rough surfaces, smooth surfaces, as well as polymerlike structures. Porous materials can be made from all these synthetic precursors and the structure of the final product reflects the geometry of the precursors. In addition, porous materials can be made by a phase-separation/leaching process which gives rise to a completely different type of porosity network. Structures from these various classes have been studied by small-angle x-ray and neutron scattering and in some cases it is possible to directly trace the structure of the porous material to the geometry of the precursor macromolecules.

INTRODUCTION

Of the many complex issues related to porosity, the understanding of its origin in both synthetic and natural materials has received the least attention. The purpose of this paper is to characterize porosity in synthetic materials and to explain the observed porosity in terms of the chemical and physical processes in precursor phases. We also wish to identify some general principles which can be used to tailor the structure of porous materials to meet structural goals.
To understand the origin of porosity, it is first necessary to characterize it. In the past, simple models based on bubbles, cylindrical channels, or packed grains have been used to predict the adsorption and flow properties of porous materials. Although these models may adequately represent experimental data, they are often inadequate when the issue is structure itself.

---

*This work performed at Sandia National Laboratories, Albuquerque, NM and supported by the U.S. Department of Energy under Contract No. DE-AC-04-76DP00789, for the Office of Basic Energy Sciences, Division of Materials Science.

We concentrate on models described by fractal geometry because simple random growth processes often lead to fractal structures. In many cases, these structures are observed in porous materials. Furthermore, since the chemical and physical conditions which control structure are very complex, it is necessary to isolate the essential factors which control structure. Even though the concept of fractal geometry is highly simplified, it does provide the means for identifying these essential factors.

Figures 1 and 2 compare schematic 2-dimensional representations of fractal and non-fractal porosity. Figure 1a is a quasi-monodisperse distribution of bubbles, whereas Fig. 1b is a loosely packed quasi-monodisperse array of grains. Neither of these materials is fractal because each has a characteristic length, the mean size of the bubbles or grains. Figure 2 shows three different classes of fractally porous materials. Fractal materials display dilation symmetry in the sense that the structure is invariant with scale changes (such as a change in the magnification of a microscope) so, within limits, they have no characteristic length. Figure 2a, for example, is a porous object with fractally rough surfaces. A magnified version of the short-length scale roughness looks exactly like the large-scale roughness. The same pattern occurs for the power-law distributed pores in Fig. 2b. Here there are pores of every size scale so that, if one zooms in, the structure does not change its basic geometrical features. Finally, Fig. 2c represents a random network which, although qualitatively different from Figs. 2a and 2b, also displays a range of scale-invariant geometry.

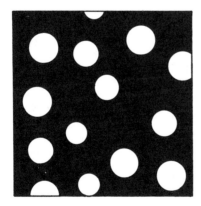

Fig. 1   Non-fractal porous materials in two dimensions:
         a). bubbles; b). packed grains.

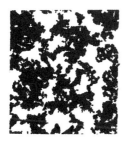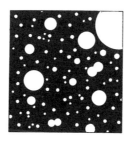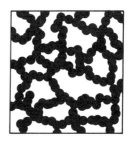

Fig. 2   Schematic representation of three 2-dimensional models for fractal porosity: a). fractally rough porous surfaces; b). power-law distribution of smooth pores; and c). network of colloidal particles.

Fractal geometry provides a quantitative measure of disorder and thus permits characterization of random systems including porous materials, polymers, and colloidal aggregates. For the case of porous materials, we anticipate fractal structures in situations where the precursor chemistry and physics involves random growth processes such as diffusion-limited aggregation or random polymerization.

Figures 1 and 2 are 2-dimensional representations of structure. These simple patterns are usually not observed in 3-dimensional materials because typical real-space observations involve a projection of a 3-dimensional object onto a plane. To study the nature of 3-d objects we exploit the techniques of small-angle x-ray and neutron scattering. These methods are sensitive to bulk properties and do not compromise the interpretation of structure by the projection mentioned above. In small-angle techniques, a beam of radiation (typically x-rays or neutrons) impinges upon the sample and one measures the intensity, I, scattered as a function of angle. In order to probe structures on the 10 to 100 Å lengthscale one measures the scattering at very small angles, typically less than one degree.

SMALL-ANGLE SCATTERING

Many random porous material show overall scattering curves which are similar in shape[1,2]. A typical scattering profile is shown in Fig. 3. Four regimes are delineated as a function of the momentum transfer K ($K = 4\pi\lambda^{-1} \sin \frac{\theta}{2}$, θ equals the scattering angle and λ is the radiation wavelength). Each of the regimes corresponds to the structure on a different lengthscale. In the limiting regime ($K \to 0$), one is probing very long wavelength fluctuations (K^{-1}). In the dilute limit, the scattered intensity in this limiting regime near $\theta = 0$ is proportional to the number density, ρ, of scatterers (pores or particles) times the square of their volume, V. In the Guinier

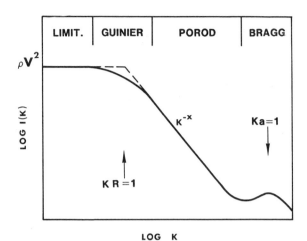

Fig. 3 Generalized scattering from disordered materials. R is the mean pore size and a is the bond length or lattice constant.

regime, one probes fluctuations comparable to the correlation range, which is typically the mean pore size. The curvature in the Guinier regime, therefore, provides the mean pore radius, R. At larger K, in the Porod regime, one measures fluctuations which are small compared to the size of the pores, but large compared to the atomic structure of the material. For fractal materials we always observe power-law behavior in the Porod regime

$$I(K) \sim K^{-x} \qquad (1)$$

Finally, in the Bragg regime, one observes structure on the atomic scale. Information on short range atomic correlations are inferred from the scattering profile in the Bragg regime. (Note that all scattering takes place at interfaces, so a hole in a porous material scatters like a particle of the same shape in solution).

From the above analysis of the scattering curve, it is possible to derive the exponent x in Eq. 1, which describes the curve in the Porod regime. In order for the curves to match smoothly in the Guinier regime, the scattered intensity must have the following general form in the Porod regime,

$$I(K) \sim \rho V^2 (KR)^{-x} \sim \rho S \qquad (2)$$

Here S is the surface area per void (particle). Note that Eq. 2 has the right limiting behavior. When $KR = 1$, it is equal to ρV^2, as expected, and it is indeed power-law. To calculate x, we exploit the fact that all scattering takes place from interfaces so in the Porod regime the scattered intensity should be proportional to the specific surface, ρS, as indicated on the right side of Eq. 2. Fractal geometry can relate V, R, and S, and therefore fix the exponent x.

For fractally rough pores, such as in Fig. 2a, $V \sim R^d$, where d is the dimension of space (typically 3) and R is the pore size. Also, if the pores are fractally rough[3]

$$S \sim R^{D_s} \quad (3)$$

where D_s is the surface fractal dimension. Substituting these relations into Eq. 2

$$I(K) \sim K^{D_s-6} \quad (4)$$

The exponent D_s-6 given in Eq. 4 was originally obtained by Bale and Schmidt[4], although the procedure used here follows Martin[5]. Note that for smooth surface porosity, $D_s = 2$ and the scattering curve decays with the slope of -4. This result is called Porod's law and is the signature of scattering from sharp interfaces. In general, however, D_s need not equal 2 and is expected to lie between 2 and 3 for fractally rough surfaces. The signature of a fractally rough surface, therefore, is a scattering law whose slope lies between -3 and -4. Occasionally scattering curves are observed with power-law slopes less than -4. In the past these have been interpreted in terms of scattering from a uniformly diffuse interface[6]. The above analysis, however, suggests that scattering laws with asymptotic (large K) slopes less than -4, correspond to D_s less than 2. Although $D_s < 2$ asymptotically ($R \to \infty$) corresponds to a disconnected surface, we see no reason why Eq. 3, with $D_s < 2$, could not describe the R-dependence of S over a limited range of length scales. We find behavior consistent with this notion for surfactant coated particles and interpret the scattering curve in terms of a "subfractally rough" interface. Physically these subfractal interfaces become smooth[7] at large length scales so scattering curves with power-law slopes less than -4 are expected only at large K.

Within the analysis presented here we do not distinguish between self-affine and self-similar surfaces. Self-affine surfaces are single-valued and do not contain overhangs whereas self-similar surfaces do. In terms of scattering, however, the only factor which is important is contained in Eq. 3 which relates the surface area to size. In fact, even a power-law distribution of smooth particles (Fig. 2b) can give rise to surface-area relations like Eq. 3. If the porosity is power-law distributed[3,4], for example, then Eqs. 3 and 4 hold with $D_s = \gamma + 1$, where γ describes the pore size distribution, $P(R) \sim R^{-\gamma}$.

We now consider the case of chain-like or mass fractal porosity shown in Fig. 2c. In this case the surface area is proportional to the volume, and both scale with pore size, R, in a fractal manner

$$S \sim V \sim R^D \qquad (5)$$

where D is the mass-fractal dimension which lies in the interval $1 < D < 3$. Substituting these relations into Eq. 2 shows that the exponent is -D:

$$I(K) \sim K^{-D}, \quad 1 < D < 3 \qquad (6)$$

For mass-fractal objects the power-law slope is greater than -3, so one can distinguish[2] between surface-fractal porosity and mass-fractal porosity by noting whether the slope is less than or greater than -3. If the observed slope is exactly -3 the structure is so rough that $D_s = D = 3$ and both the surface and the volume scale with the dimension of space.

If one imagines starting with porosity which is smooth, such a structure should give an intensity profile with a slope of -4 (Porod's law). As the surface becomes increasingly rough, in a fractal way, the slope changes from -4 toward -3. When the slope of -3 is reached the object is, in a sense, totally rough and can be considered either a surface or mass fractal. This point is the crossover to a mass-fractal structure, which gives a scattering profile > -3.

The analysis presented here differs from that of Rojanski et al.[8], who believe that the situation $D_s = D = 3$ gives rise to scattering profiles with a slope of -4, identical to Porod's law. In our opinion, scattering profiles pass smoothly from the surface-fractal to mass-fractal regime[9]. The Rojanski analysis, on the other hand, depends on a particular unique choice for the cut-off function which is, in general, not justified.

Fractal analysis clearly does not apply to all materials. Scattering from a porous solid created by packing of uniform grains (Fig. 1b), for example, is not fractal and, therefore, would not give rise to power-law profiles. For quasimonodisperse grains it is reasonable to expect a peak in the scattering curve when the K is comparable to the inverse particle diameter. This peak is analogous to that of observed in simple liquids.

Having established some general rules for the interpretation of scattering curves, we now consider the structure of synthetic silicas prepared by different processes. Some of these materials fall neatly into the categories described above, whereas others do not. The goal is to establish structure from the scattering curves and then identify the essential factors that control structure in the precursor phases.

SILICA AEROGELS

Aerogels from Colloidal Precursors: We first considered a series of commercial aerogels of varying densities. These materials were obtained from Airglass AB, Sjobo, Sweden. The materials are prepared by base-catalyzed hydrolysis and condensation of tetramethyl-orthosilicate (TMOS) in alcohol[10]. Polymerization ultimately leads to a gel which is critical-point dried to remove the solvent. The resulting porous solid is nearly transparent. The materials used in this experiment varied in density from .09 to .21 g/cm^3. The data for these materials are shown in Fig. 4[11].

The only feature of the scattering profiles in Fig. 4 which appears in all three data sets is the limiting slope of -4 at large K. As indicated above, this slope is the signature of smooth surfaces. For the data in Fig. 4 we can say that these smooth surfaces exist on dimensional scales of the order of 10 Å. It is the presence of these smooth surfaces that lead us to believe that these structures are colloidal or particulate in nature. For the case of the lowest density aerogel, a second power-law regime is observed corresponding to dimensions of about 50 Å.

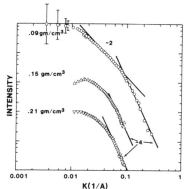

Fig. 4 SAXS profiles for three silica aerogels prepared from TMOS under base-catalyzed conditions. W=4.0.

In this case the scattering profile has a slope $\simeq$ -2. This slope corresponds to a mass-fractal object and suggests that this material is similar to that shown in Fig. 2c. That is, it is a chainlike arrangement of colloidal particles which form an open network. The fractal dimension of the network backbone is 2, similar to that of colloidal silica aggregates observed in solution[12].

The origin of the network structure inferred for the low density aerogel can be found in the precursor chemistry of this material. These materials are prepared from TMOS under base-catalyzed conditions with excess water. Under these conditions, condensation polymerization proceeds by monomers attaching to clusters, a situation which leads to uniform colloidal particles (see below). Once the colloidal particles are formed, they aggregate by a conventional colloid aggregation process which leads to the network structure shown in Fig. 2c. The structure of the porous solid, then reflects both the polymerization and colloid aggregation processes in the solution precursor.

For the higher density aerogels, there is no evidence for a fractal network. In these cases a peak is observed in the scattering profile. We interpret the peak in the scattering curve much like that from observed in the wide-angle pattern for an atomic liquid where peak represents the mean distance between atoms. For these higher density aerogels the conventional picture of loosely packed spheres (Fig. 1b) appears to be qualitatively correct. Although colloid aggregation may occur, the systems are not fractal over a significant range of length scales.

<u>Xerogels from Rough Colloidal Precursors</u>: A slight modification in the precursor silica chemistry leads to a qualitatively different type of porous material[9]. For this second system we use tetraethylorthosilicate (TEOS) as opposed to TMOS but still polymerize under base-catalyzed conditions. In addition, the hydrolysis takes place at lower water to silicon ratio, $W=[H_2O]/[Si]$.

Although the chemistry is similar to the TMOS system in Fig. 4, hydrolysis is substantially less aggressive so that the polymerizing species in general is not silicic acid, $Si(OH)_4$, but consists of a mixture of silicates of lower functionality. Subsequently, we show that these conditions produce fractally rough colloidal particles.

Figure 5 shows the scattering curves for polymerized TEOS macromolecules in solution and for a solid made by air drying the solution. The measured slopes of -3.2 and -3.3 indicate fractally rough surfaces. In the solution, we envision colloidal particles with uniform porous interiors and fractally rough surfaces. The solid then consists of a loose packing of these rough particles. Figure 6 schematically illustrates the proposed structure of the solution and the solid.

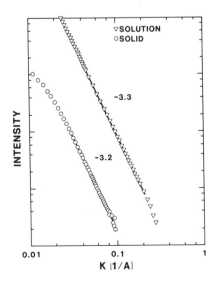

Fig. 5 Scattering profiles for silicate polymers grown from TEOS in alkaline solution (triangles) and porous solid made by air drying the solution (circles). Initial concentration of NH_4OH is .01 M, and the water ratio, W=2.0. Solution data were taken eight days after the initiation of the reaction.

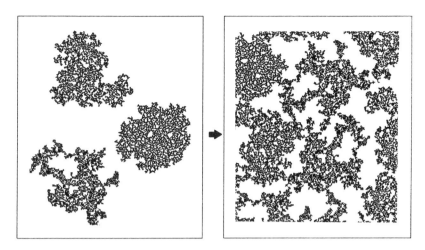

Fig. 6 Schematic illustration of fractally rough particles in solution (a) and packed to form a porous solid (b).

Aerogels from Polymeric Precursors: Recently Schaefer and Keefer demonstrated the existence of chainlike silica polymers in solution[13]. They showed that under a two-step polymerization process, ramified, mass-fractal objects are produced on lengthscales between 10 and 100 Å. Later these authors identified the two-stage polymerization process as the key to the production of non-colloidal silicates[14]. We now describe a porous solid made by a two-stage process and also find mass-fractal structures.

Figure 7 shows the scattering profile for an aerogel prepared from a two-stage precursor. In the first stage, TEOS is dissolved in ethanol and polymerized under acid catalysis with a water/TEOS ratio W = 1.3. This solution is refluxed for several hours. Then excess alcohol is distilled off and the product is heated to 120° under vacuum. The resulting polymeric oil has a number-average molecular weight of 2900. Elemental analysis of this oil is consistent with substantial ladder polymer content. In the second stage, the polymeric oil is dissolved in butanol and gelled under basic conditions by the addition of water to a final concentration W=2.5.

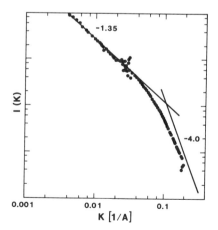

Fig. 7 Scattering profile for a polymeric aerogel prepared in a two-stage polymerization process.

The resulting gel is then supercritically extracted to yield the aerogel reported in Fig. 7. The power-law region of this curve has a slope of -1.35 indicating a mass-fractal structure. Although there is deviation from power-law behavior near $K = .1 \text{ A}^{-1}$, the slope never reaches -4, the value expected for colloidal subunits.

Based partially on the data in Fig. 7, we believe that the structure of this material is a branched ladder polymer[15]. The fact that the slope in the fractal region is less than that observed for the polymeric silicates studied by Schaefer and Keefer[13], indicates that this structure is somewhat more expanded than a randomly branched polymer. This behavior would be expected for stiff ladder-polymer chains. In addition, if the backbone consisted of single-strand polymers, one would anticipate that the structure would collapse upon drying. The structural rigidity of a ladder polymer, however, provides the necessary strength to support drying of the aerogel. Finally, the crossover to more compact behavior observed near $K = .1 \text{ A}^{-1}$ is consistent with a ladder-polymer backbone.

The data contained in Figs. 4, 5, and 7 indicate that porous silicas can be prepared with different classes of porosity. We have demonstrated the existence of structures formed from smooth colloidal particles, rough colloidal particles, and polymeric precursors. We now wish to identify the essential factors which control growth in the solution precursors and, therefore control the structure of the final porous object.

Precursor Growth and Structure: The important factors which control structure in solution are pH, water ratio, W, and the reaction sequence. Clearly there is a complicated interplay of factors which make prediction of structure based on these factors difficult. Nevertheless, by mapping polymerization onto simple physical models, we identify some essential factors which relate structure to solution chemistry[14].

Figure 8 shows four basic growth models which have been extensively studied by computer simulation[16]. These models are distinguished by the factors listed on the axes: monomer-cluster growth versus cluster-cluster growth, and diffusion-limited versus reaction-limited growth. The Eden model, for example, represents reaction-limited monomer-cluster growth. Eden growth produces compact clusters with asymptotically smooth surfaces. By contrast, diffusion-limited cluster-cluster (DLCA) growth produces open ramified structures. Since diffusion-limited reactions are not found in the silicate system we ignore the right column. We concentrate on the two reaction-limited models, so called Eden growth (reaction-limited monomer-cluster) and reaction-limited cluster aggregation (RLCA). From the left column of Fig. 8 we conclude that compact clusters are grown under monomer-cluster growth conditions, whereas ramified clusters (mass-fractals) are grown under cluster-cluster growth conditions. We hope to map the polymerization chemistry onto these two models.

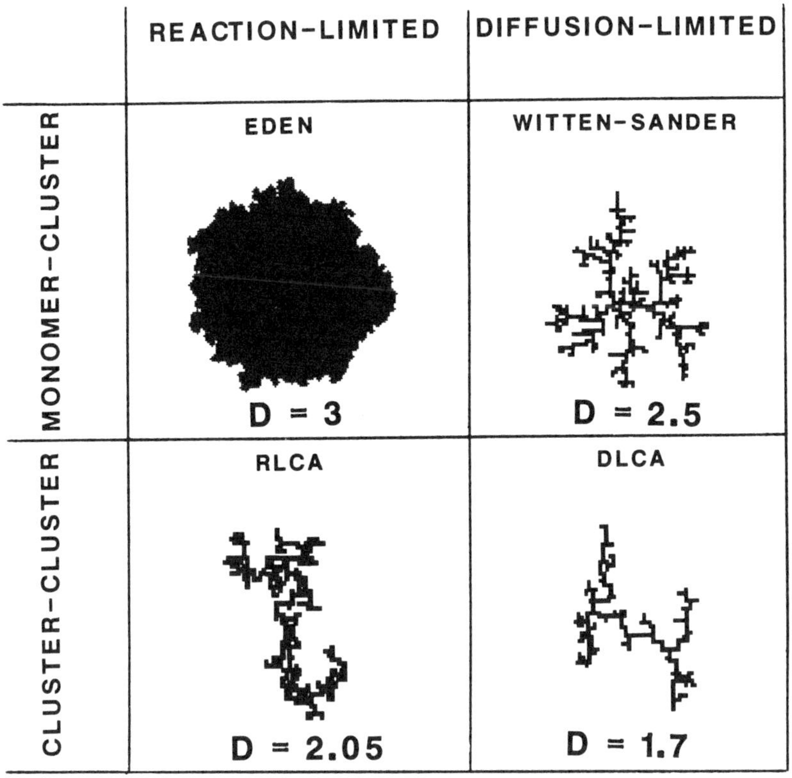

	REACTION-LIMITED	DIFFUSION-LIMITED
MONOMER-CLUSTER	EDEN D = 3	WITTEN-SANDER D = 2.5
CLUSTER-CLUSTER	RLCA D = 2.05	DLCA D = 1.7

Fig. 8   Four models of random growth.

Solution pH effects polymerization through two factors; it modifies the rates of the hydrolysis and condensation reactions and it controls the reaction mechanisms. In the sol-gel systems described above, two reactions are involved in the solution precursors. In the first, functional sites, OH-groups, are generated by the hydrolysis of non-condensable alkoxide groups:

$$Si(OC_2H_5)_{4-(n-1)}(OH)_{n-1} + H_2O \rightarrow Si(OC_2H_5)_{4-n}(OH)_n + C_2H_5OH; \quad 1 \leq n \leq 4. \quad (7)$$

The silicate species on the righthand side of Eq. 7 is a "n-functional monomer" which can form n Si-O-Si linkages by condensation of the hydrolyzed species.

$$R_k Si(OH) + R_j Si(OH) \rightarrow R_k Si-O-SiR_j + H_2O. \quad (8)$$

R_k is either -OH or -OC$_2$H$_5$.

The effect of pH on rates of Eqs. 7 and 8 is illustrated schematically in Fig. 9, which shows the relative rates of the hydrolysis and condensation reactions as a function of pH. Note that at pH, around 2, the condensation reaction is the limiting step, whereas under base-catalyzed conditions, around pH 9, hydrolysis is the limiting step[17].

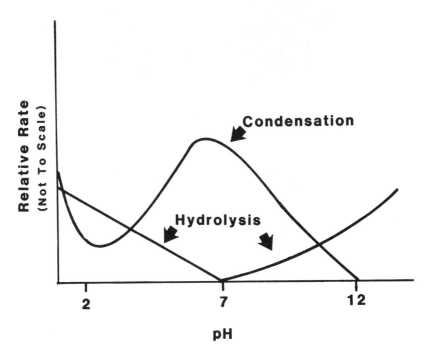

Fig. 9   Approximate pH dependence of the hydrolysis and condensation rates for silicates.

Under base-catalyzed conditions, monomer-cluster growth is favored because the rapid condensation reaction will consume monomers as they are produced. Under acid-catalyzed conditions, however, hydrolysis is rapid[17] and produces a burst of monomers which then slowly condense by cluster-cluster growth. Although neither class of growth is excluded by these arguments, we conclude that base-catalyzed conditions favor monomer-cluster growth, whereas acid-catalyzed conditions favor cluster-cluster growth. By reference to Fig. 8 we conclude that mass-fractals are favored under acid-catalyzed conditions, whereas uniformly dense clusters are favored under base-catalyzed conditions[14].

The pH also effects growth through the reaction mechanism. This factor has been analyzed by Keefer[18], who observes that under base-catalysis, hydrolysis and condensation take place by nucleophilic substitution. Because this mechanism requires inversion of one of the reacting species, base-catalysis biases the system toward monomer-cluster growth. This bias occurs because one of the reacting species (the monomer) can easily invert[9]. Cluster-cluster reactions are inhibited because neither reacting species can invert.

We conclude that both the reaction kinetics and the reaction mechanism favor growth from monomers. Except when W is small, therefore, we expect that chainlike mass-fractal objects are excluded under base-catalyzed conditions.

The above analysis explains the structure of the two porous solids produced under base-catalyzed conditions. The short-range structure of the aerogels in Fig. 4 is consistent with the Eden model which yields compact clusters with smooth surfaces. These aerogels were prepared under base-catalyzed conditions with excess water, precisely the conditions which lead monomer-cluster reaction-limited growth from fully hydrolyzed monomers (n=4 in reaction 7). The network structure on larger scales is due to colloid aggregation.

The materials in Fig. 5 are produced with less aggressive hydrolysis conditions and yield structures which, although uniform, have fractally rough surfaces. We explain the structures of these materials by a variation of the Eden model called poisoned Eden growth. In this case, monomers are not necessarily fully hydrolyzed because of the mild hydrolysis conditions (n < 4). Unhydrolyzed sites are poisoned and precluded from growth. We have simulated Eden growth from poisoned monomers and find that the poisoned Eden model produces uniform structures with fractally rough surfaces. We call these macromolecules fractally rough colloids[9,19]. The porous solid in Fig. 5 is produced by drying and loosely packing rough colloids as illustrated in Fig. 6.

The above analysis indicates that high pH is the essential factor which favors compact growth. Secondary factors such as the water ratio and the relative speed of the hydrolysis reaction determine whether the particles are smooth or fractally rough.

We now consider the modifications of chemistry needed to produce polymeric structures[14]. The polymeric aerogel shown in Fig. 7 is produced by a two-stage polymerization process. The first stage is acid-catalyzed and the second is base-catalyzed. Both the acid catalysis and the two-stage reaction sequence favor ramified structures by biasing the system toward cluster-cluster growth.

First we consider the effect of acid catalysis. Under these conditions, hydrolysis is rapid compared to condensation (see Fig. 9) so a burst of reactive-monomers is produced which then slowly condense via the cluster-cluster path. Also, the electrophilic mechanism[18] of the acid-catalyzed hydrolysis tends not to produce the orthosilicic acid monomer which is favored under basic conditions. Both the reaction kinetics, which allow cluster-cluster reactions, and the mechanism, which favors n < 4, bias the system toward ramified polymers and away from compact colloids.

The above arguments for ramified structures in acidic solution are fortified by the two-stage reaction sequence[14]. In the first stage, a burst of monomers is produced with most monomers having at least one condensable hydroxyl group. After the first stage, the system consists of small clusters. In the second stage, further growth must necessarily proceed by cluster-cluster growth since all the monomers are depleted in the first stage. Regardless of catalysis, ramified polymers will result from the reaction-limited cluster-cluster growth in the second stage.

The three factors described above all favor the growth of polymeric structures under acid-catalyzed two-stage polymerization. It is not surprising then that the scattering curves for porous solids prepared in this way show a very low slope (-1.35) consistent with a mass-fractal backbone. We describe these materials as polymeric aerogels.

LEACHED BOROSILICATE GLASSES

In the previous section we concentrated on porous materials formed by the so-called sol-gel process. In this section we discuss a different class of porous glasses: porous silica prepared by leaching of a phase-separated precursor[20]. These materials are prepared by quenching of a borosilicate glass from the melt and subsequently removing the borate phase by acid dissolution.

Figure 10 shows the scattering curve for commercial leached borosilicate glass, Corning Vycor 7930. The scattering curve has a slope of -4 in a large-K region indicating that the surfaces of the pores are smooth. In addition, a peak is observed in the scattering profile similar to that found in the high density aerogels in Fig. 4.

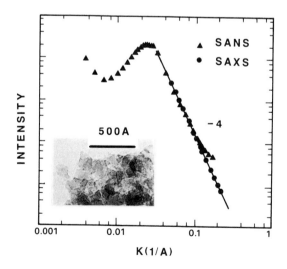

Fig. 10 Scattering curves for Vycor 7930. Deviation of the SANS data from Porod's law near $K = .2$ Å^{-1} is due to the existence of an uncorrected incoherent background. Scattering curves are corrected for the empty beam background intensity. The inset is a TEM of the sample.

In the present case, however, we explain the peak in terms of the phase separation process. During the quench from the melt, phase separation takes place by the spinodal process which favors the growth of a specific Fourier component in the concentration fluctuation spectrum[21]. It is well known that spinodal decomposition results in a peak in the scattering curve at the K corresponding to this unstable Fourier frequency. The peak in Fig. 10 presumably represents the frozen-in unstable Fourier component.

PACKED ALUMINA POWDERS

The final porous system we consider is prepared from alumina powders. We study both the dry powders and a fragile porous solid prepared by suspending the powder in a surfactant solution and drying the suspension. The surfactant used in the suspension procedure is sodium dodecyl sulfate. The powder used is a commercial boehmite (AlOOH, Alcoa Monal 100). The nominal surface area of the powder is 100 m^2/g.

Figure 11 shows the SAXS curve for both the powder and dried suspension. The powder data show normal Porod law scattering in the high angle region indicating that the particles have smooth surfaces. The dried suspension, however, shows a slope more negative than -4. Based on Eq. 3 this slope corresponds to a subfractal surface with D_s=1.5.

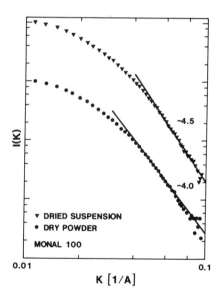

Presumably the negative deviation from Porod's law is caused by the presence of the surfactant at the interface between the particle and pore. Within the subfractal interpretation this surfactant leads to a surface structure whose roughness decreases with distance. A schematic representation of the rough interface idea is presented in Fig. 12a.

Fig. 11 SAXS data for suspended and packed alumina powders.

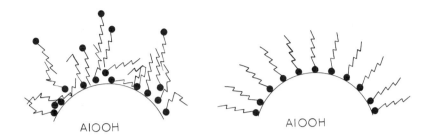

Fig. 12 Models for a surfactant coated interface: a). subfractally diffuse and b). uniformly diffuse.

Negative deviations from Porod's law can also be interpreted in terms of a uniformly diffuse interface as illustrated in Fig. 12b. The interface width can be obtained[22] from a plot of $K^4 I(K)$ versus K^2 as illustrated in Fig. 13. The calculated interface width is only 1.06 Å. This value is an order of magnitude smaller than the length of the SDS molecule, suggesting either that the molecules are lying flat on the surface (unlikely) or that the uniformly diffuse interface interpretation is incorrect. Our prejudice therefore is to accept the subfractal interpretation of the negative deviation from Porod's law. Similar negative deviations were observed by Schaefer and Hurd[23] for dried suspensions of surfactant coated carbon black powders.

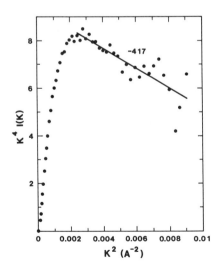

Fig. 13 Analysis of dried alumina powder assuming a diffuse interface.

CONCLUSION

From the scattering studies presented here, we conclude that a variety of porous materials can be prepared. Although we find evidence in some cases for the conventional picture of packed grains, this model does not describe the structure of most porous materials studied. Porous silicas may contain fractally rough pores if they are formed by the compaction of rough particles. In other cases, structure is determined by a colloid aggregation process in the solution precursor. Polymeric aerogels also exist, these being prepared by a two-stage polymerization process. Finally, for phase separated systems, the physics of phase separation control the ultimate structure of the porous material.

Since the geometry of porosity must play an important role in many physical properties of xerogels, we believe the ideas presented here may help interpret existing data in terms of structure. In addition, manipulation of the simple factors we have identified allow one to tailor a structure for a specific chemical or physical goal.

ACKNOWLEDGEMENT

We thank Alan Hurd for suspending the alumina particles. We thank Steve Spooner who supervised the data collection at the National Center for Small Angle Scattering Research.

REFERENCES

1. D. W. Schaefer and K. D. Keefer, Mat. Res. Soc. Symp. Proc. $\underline{32}$, 1 (1984).
2. D. W. Schaefer, J. E. Martin, A. J. Hurd, and K. D. Keefer in Physics of Finely Divided Matter, M. Boccara and M. Daoud, Eds. (Springer Verlag, New York, 1985) p. 31.
3. P. Pfeifer and D. Avnir, J. Chem. Phys. $\underline{79}$, 3558 (1983).
4. H. D. Bale and P. W. Schmidt, Phys. Rev. Lett. $\underline{53}$, 596 (1984).
5. J. E. Martin, J. Appl. Cryst. $\underline{19}$, 25 (1986).
6. W. Ruland, J. Appl. Cryst. $\underline{4}$, 70 (1971).
7. P.-Z. Wong, Phys. Rev. B $\underline{32}$, 7417 (1985).
8. D. Rojanski, D. Huppert, H. D. Bale, X. Dacai, P. W. Schmidt, D. Farin, A. Seri-Levy, and D. Avnir, Phys. Rev. Lett. $\underline{56}$, 2505 (1986).
9. K. D. Keefer and D. W. Schaefer, Phys. Rev. Lett. $\underline{56}$, 2376 (1986).
10. S. Henning and L. Svensson, Phys. Scr. $\underline{23}$, 697 (1981).
11. D. W. Schaefer and K. D. Keefer, Phys. Rev. Lett. $\underline{56}$, 2199 (1986).
12. D. W. Schaefer, J. E. Martin, P. Wiltzius, and D. S. Cannell, Phys. Rev. Lett. $\underline{52}$, 2371 (1984).
13. D. W. Schaefer and K. D. Keefer, Phys. Rev. Lett. $\underline{33}$, 1383 (1984).

14. D. W. Schaefer and K. D. Keefer, Mat. Res. Soc. Symp. Proc. 73, 295 (1986).
15. T. E. Helminiak and G. C. Berry, J. Pol. Sci. Pol. Symp. 65, 107 (1978).
16. H. E. Stanley and N. Ostrowsky, *On Growth and Form* (Martinus-Nijhoff, Boston, 1986).
17. R. A. Assink and B. D. Kay, Mat. Res. Soc. Symp. Proc. 32, 301 (1984).
18. K. D. Keefer, Mat. Res. Soc. Symp. Proc. 32, 15 (1984).
19. K. D. Keefer, Mat. Res. Soc. Symp. Proc. 73, 277 (1986).
20. D. W. Schaefer, J. P. Wilcoxon, and B. C. Bunker, Phys. Rev. Lett. XX, XXX (1986).
21. J. W. Cahn, Trans. Met. Soc. AIME 242, 166 (1968).
22. J. T. Koberstein, B. Morra, and R. S. Stein, J. Appl. Cryst. 13, 34 (1980).
23. D. W. Schaefer and A. J. Hurd, Proc. Electrochem. Soc. 85-8, 54 (1985).

II. Fluid Flow

INCORPORATING THE INFLUENCE OF WETTABILITY INTO MODELS OF IMMISCIBLE FLUID DISPLACEMENT THROUGH A POROUS MEDIA

E. B. Dussan V.
Schlumberger-Doll Research
Old Quarry Road
Ridgefield, CT 06877-4108

ABSTRACT

One approach of assessing the influence of capillary forces on immiscible fluid displacement through porous media is to perform an analysis of the fluids on a length scale consisting of several pores. In order to perform such an analysis, a model must first be identified. This consists of specifying: the geometric structure of the pores and their interconnections; the material properties of the two immiscible fluids, e.g., whether or not they are Newtonian, the values of their viscosities and their interfacial tensions, and whether or not surface-active-agents are present; and, the wettability of the system. The objective of this presentation is to indicate a procedure by which wettability can be incorporated into such an analysis. The procedure is well-defined and straight forward, containing only experimentally measurable properties of the system despite the current lack of understanding of the physics by which one fluid displaces another fluid across the surface of a solid.

In section 1 the current status of dynamic wettability is reviewed from the perspective of fluid mechanics. A well-posed boundary-value problem whose solution describes the dynamics of a meniscus containing a contact line is identified in section 2. An example is presented in section 3. A summary and conclusions appear in section 4.

1. DYNAMIC WETTABILITY FROM THE PERSPECTIVE OF FLUID MECHANICS

It is instructive to begin by identifying the mechanism by which wettability influences the state of an entire fluid body. Wettability refers to the relative strengths of the interactions of each fluid with the solid. Thus, it addresses the state of the material in the immediate vicinity of the contact line, i.e., at the intersection of the fluid-fluid interface and the surface of the solid. When stated in these terms, it is somewhat surprising that such a small amount of material can have a substantial influence on the state of the entire body of fluid. The validity of this statement can be illustrated by examining the well-known problem of determining the height of rise of liquid, h, in a vertical capillary when its lower end is submerged into a dish filled with liquid; refer to figure 1.

The determination of h is simple. Two expressions can be identified for the difference between the atmospheric pressure, P_0, and the pressure in the liquid just beneath the fluid interface, P_1. They are: $P_0 - P_1 = 2\sigma/R$, where R denotes the radius of the segment of the sphere closely approximating the shape of the gas-liquid interface, and σ denotes the surface tension; and $P_0 - P_1 = \rho g h$, where we have used the fact that the pressure of the liquid at the base of the capillary under static conditions is given by P_0, ρ denotes the density of the liquid, and g denotes the gravitational constant. Equating these two expressions, and replacing R by $a/\cos\theta$ gives

$$h = \frac{2\sigma \cos \theta}{\rho g a} \tag{1.1}$$

where θ denotes the value of the contact angle.

In the above example the macroscopic consequences of the intermolecular interactions between the gas, liquid, and solid in the immediate vicinity of the contact line are evident: the intermolecular interactions determine the value of the contact angle; and, the contact angle dictates the value of h, as expressed in (1.1). Examining the relationship between h and θ in more detail gives the following chain of interdependencies: the intermolecular forces at the contact line determine the contact angle, θ, which in turn plays an important role in determining the curvature of the fluid interface, $2/R$, and the difference in pressures within the gas and the liquid across the interface, $P_0 - P_1$, which finally dictates the height of the liquid, h. Thus, we see the pivotal role played by the contact angle. The contact angle is the continuum property of all three materials in the system by which the wettability affects the fluid mechanics. If it is possible to establish in a particular situation the lack of importance of the contact angle, then wettability is not a issue.

There is good reason to anticipate that the above stated "chain of interdependencies" also applies under dynamic conditions. While the equations governing the pressure field and shape of the fluid-fluid interface are significantly influenced by the forces arising from the dynamics of the fluids, *the dependence of the shape of the interface on capillarity remains unchanged*. Thus, it seems reasonable to conclude that the *dynamic* contact angle is the appropriate boundary condition for the shape of the interface under dynamic conditions. However, establishing the validity of this assertion is by no means straight-forward, a consequence of the appearance of a singularity at the moving contact line in any analysis based on the usual hydrodynamical assumptions (Newtonian viscous incompressible fluids obeying the no-slip boundary condition at the solid surface). Simply put, these analyses are characterized by a stress tensor which asymptotes to $1/r$ as $r \to 0$, where r denotes the shortest distance to the moving contact line. This implies that the fluids exert an infinite drag force on the solid (in other words, a relationship cannot be derived between the volumetric flow rate and the pressure drop which produces it), and that the dynamic contact angle cannot be used as a boundary condition for the equation governing the shape of the fluid-fluid interface; refer to Dussan V. (1979)[1] for a detailed discussion. Suffice it to say that the presence of the singularity at the moving contact line cannot be ignored. The only way of eliminating the singularity is by changing at least one of the usual hydrodynamic assumptions referred to above. Thus, whether or not the specification of the dynamic contact angle is an appropriate boundary condition for the shape of the interface, strictly speaking, must be viewed as an assumption. Indeed, it is assumed throughout the remainder of the presentation that the specification of the dynamic contact angle *is* the appropriate boundary condition.

Two models of the behavior of the fluids in the immediate vicinity of the moving contact line have enjoyed a good measure of popularity over the past decade. One approach assumes that a very thin film of advancing fluid spreads forward over the solid surface in front of the *apparent* location of the moving contact line (*apparent* refers to the location as perceived from observation with a low power microscope). The existence of such films have been established experimentally in certain material systems, and under specified conditions. The dynamics of the fluids within these films has been analyzed by including the long range van der Waal attractive force as a body force in the Navier-Stokes equation. Unfortunately such analyses are characterized by a non-ignorable singularity at the leading edge of the thin film. Thus, the inclusion of a thin film changes the location of the singularity, but does not eliminate it; refer to de Gennes (1985)[2] for a review of this approach. It is worth noting that it has been established that this singularity cannot be eliminated by the mere inclusion of a conservative body force in the Navier-Stokes equation (Dussan V. & Davis 1974).[3] The other approach abandons the no-slip boundary condition at the solid surface in the immediate vicinity of the moving contact line. A wide range of theoretical models has been investigated predominately under conditions corresponding to vanishingly small Reynolds number and small capillary number. All of these solutions have the same basic structure in that the fluids can be segmented into two regions: *the inner region*, located in the immediate vicinity of the contact line within which the dynamics of the fluids are very sensitive to the model specified for the slip boundary condition; and, *the outer region*, incompassing the entire body of fluids excluding the inner region, within which the dynamics of the fluids can be accurately modeled by the usual hydrodynamic assumptions. If L_s and a denote the length scales characterizing the inner and outer regions, respectively; then the above structure arises only when $L_s/a \to 0$. Interestingly enough, the dynamics of the fluids in the outer region - on the length scale of primary concern when assessing the influence of wettability on flow through porous media - is affected only by the shape of the fluid-fluid interface as it leaves the inner region and enters the outer region (the validity of this statement applies only to lowest order in a formal asymptotic expansion in L_s/a, as $L_s/a \to 0$). Thus, the dynamics of the fluids in the outer region can be completely determined from the empirical measurements of the dependence of (R_I, θ_I) on the speed of the contact line for the specific material system of interest; refer to figure 2. A detailed presentation of this procedure along with experimentally obtained supporting evidence is given in Ngan & Dussan V. (1987).[4] Note that θ_I plays the role in the dynamic case identical to that of the "actual" contact angle in the static case, as illustrated at the beginning of this section with the example of liquid in a vertical capillary. It is from this perspective that the influence of wettability on immiscible fluid displacement through porous media will be pursued in the remainder of this presentation.

Although extensive data of θ_I does not exist, its general behavior can readily be deduced; refer to figure 3. The most striking aspect of the functional dependence of θ_I on U, the speed of the contact line, is its discontinuity at zero contact line speed. This is a direct consequence of the fact that θ_I does not differ in value from the actual contact angle under static conditions. The nonuniqueness of the *experimentally* measured values of the static contact angle is well-documented. However, the monotonic non- decreasing functional dependence of θ_I on U, for $U \gtrless 0$, assumed in figure 3, is based on more tenuous reasoning. The difficulty lies in the differences in the definitions of θ_I and the various dynamic contact angles whose measurements are reported in the literature. Existing analytical solutions to date indicate that the experimentally measured contact angles always exceed θ_I; refer to the last paragraph in Ngan & Dussan V. (1984)[5] for further discussion. Thus, the fact that the author knows of no experimental evidence characterized by a *decreasing* functional dependence of the dynamic contact angle on U, does not necessarily imply a *non-decreasing* functional behavior of θ_I on U. On the other hand, one may wish to discount the possibility of a decreasing functional dependence of θ_I on U based upon the fact that it gives rise to anomalous predictions; refer to Murphy (1984).[6] In any case, it is important to recognize that the functional dependence of θ_I on U is a material property of the system.

2. APPROPRIATE BOUNDARY-VALUE PROBLEM GOVERNING THE FLOW OF IMMISCIBLE FLUIDS THROUGH A POROUS MEDIA ON A LENGTH SCALE CONSISTING OF SEVERAL PORES

The appropriate boundary-value problem for bodies of fluids containing moving contact lines has been identified in section 2. It consists of the usual hydrodynamic assumptions (including the no-slip boundary condition at the solid surface), along with the specification of θ_I, the angle formed by the fluid interface at a distance R_I from the perceived location of the contact line. Although the singularity at the contact line has not been eliminated, its presence can be ignored. For example, when evaluating the drag force exerted by the fluids on the solid, the singularity can be bypassed by calculating the force exerted by the fluids on the surface within the fluids at a distance R_I from the contact line (includes the force exerted by the surface tension at the fluid interface); refer to figure 4. Since our attention is restricted to flow through porous media, the boundary-value problem can be simplified considerably.

Scales can be specified, and parameters can be identified without introducing any severe restrictions. If a (appropriate length scale associated with a pore), U_0, and σ/a denote the length, velocity, and pressure scales, respectively, then: the Navier-Stokes equation introduces the Reynolds number $(\rho U_0 a/\mu)$ and the Bond number $(\rho g a^2/\sigma)$; the boundary conditions at the fluid interface introduces the capillary number $(U_0 \mu/\sigma)$ and the viscosity ratio (μ_1/μ_2); and the boundary conditions at the surface of the solid introduces $\theta_I(U)$ and R_I/a.[1]

The usual conditions associated with flow through porous media correspond to very small values of both the Reynolds and capillary numbers. So small, in fact, that the lowest order mode of an asymptotic expansion in these parameters suffice. The boundary-value problem in dimensional form for this mode is:

$$0 = -\nabla P - \rho g \underline{k} \qquad \text{within each fluid,} \qquad (2.1)$$

where the pressure datum may be time dependent;

$$\frac{\sigma}{R_M} = P_1 - P_2 \qquad \text{evaluated at the fluid interface,} \qquad (2.2)$$

where R_M denotes the mean radius of curvature of the fluid interface; and,

$$\theta = \theta_I(U) \qquad \text{at the contact line,} \qquad (2.3)$$

where θ denotes the contact angle. Thus, the movements of the meniscii, implying the volumetric flow rate through the solid structure, can be determined before the lowest order velocity of each fluid phase. Note that the singularity at the moving contact line does not appear in the boundary-value problem associated with this order mode.

---

[1] In some problems, e.g., a drop spreading on a solid, the appropriate velocity scale is obtained from $\theta_I(U)$. One such choice is $(d\theta_I/dU)^{-1}$. This should not be confused with the velocity scale σ/μ, resulting from the competition between the viscous and the capillary forces at the fluid interface appearing in the capillary number.

4. EXAMPLE

The behavior of an index of fluid as it moves through a straight circular capillary of length L and radius a connecting two reservoirs of another immiscible fluid illustrates many of the concepts addressed in the previous section; refer to figure 5. Even though the capillary and reservoirs can be regarded as an interconnection between two pores, no contention is being made that this geometric model captures all the important features of the geometric structure of pores and their interconnections in a porous media.

The analysis of the motion of the fluids, correct to lowest order, is rather straight-forward. Equation (3.1) implies that the pressure fields within the reservoirs on the left and right ends of the capillary, and within the index of fluid are at most a function of time (gravity is being neglected). These pressures are denoted by P_H, P_L, and P_D, respectively. The fact that the pressures are constant with respect to position, and (2.2), imply that the mean radius of curvature of the two interfaces can at most depend on time. Since the capillary has a circular cross-section, the shapes of the interfaces must be that of spherical caps, giving $P_H - P_D = (2\sigma/a)\cos\theta_H$ and $P_L - P_D = (2\sigma/a)\cos\theta_L$ for the left and right meniscus, respectively. Subtracting these two expressions gives

$$P_H - P_L = \frac{2\sigma}{a}\left\{\cos\theta_I(-U_D) - \cos\theta_I(U_D)\right\}, \tag{3.1}$$

where $-U_D$ and U_D denote the speed of the contact line associated with the left and right meniscus, respectively. Thus, (3.1) gives the principal result relating the volumetric flow rate through the capillary, $U_D \pi a^2$, to the difference in pressure required to produce it.

The consequences of the assumed form of $\theta_I(U)$ on the dynamic behavior of the index of fluid in the capillary can be identified. The most obvious characteristic is the necessity for

$$P_H - P_L > \frac{2\sigma}{a}\{\cos\theta_R - \cos\theta_A\} \tag{3.2}$$

when the index moves through the capillary. If (3.2) is not satisfied, then the index is stationary. Its shape is unique, to within a translation, only when the equality sign is valid. Thus, the discontinuity in θ_I at $U = 0$, often referred to as capillary hysteresis, causes bodies of fluids to appear to "stick" to the solid surfaces because the bodies can sustain a difference in pressure without moving; refer to Dussan V. (1985)[7] and (1987)[8] for other examples.

It is also seen that contact angle hysteresis and the monotonic dependence of θ_I on U are both responsible for the irreversible conversion of mechanical into nonmechanical energy. This can be illustrated by determining the work, W, performed on the index when it is forced to travel from the left end to the right end of the capillary, and back again. To lowest order,

$$W = 2\pi a \sigma \int_0^T U_D \{\cos \theta_I(-U_D(t)) - \cos \theta_I(U_D(t))\} dt, \qquad (3.3)$$

where T denotes the period of time for the motion,

$$L = \int_0^{T_L} U_D(t) dt = -\int_{T_L}^T U_D(t) dt,$$

and T_L denotes the period of time required for the index to travel from the left end to the right end of the capillary. If θ_I is a monotonically non-decreasing function of U, then $W > 0$, with its minimum value, W_{MIN}, given by

$$W_{MIN} = 4\pi a L \sigma \{\cos \theta_R - \cos \theta_A\}.$$

Thus, $d\theta_I/dU > 0$ implies a dissipation of mechanical energy; refer to Dussan V. & Davis (1986)[9] for further discussion.

It is instructive to compare the dissipation of energy associated with the moving contact line with that associated with the viscosities of the fluids. Consider the special case characterized by

$$U_D = \begin{cases} U_0 & \text{for } 0 < t \leq T/2 \\ -U_0 & \text{for } T/2 < t \leq T \end{cases},$$

where U_0 is a constant; and by fluids with the same viscosity. If the viscous dissipation of the fluids is approximated by that associated with Poiseuille flow, then

$$\frac{Viscous\ Dissipation}{\substack{Dissipation\ Associated\ With \\ Moving\ Contact\ Line}} = \frac{\frac{L}{a}\frac{U_0\mu}{\sigma}}{\cos\theta_I(-U_0) - \cos\theta_I(U_0)}$$

Refer to Table 1 for an evaluation of the function $\cos\theta_L - \cos\theta_R$. Thus, if $\theta_I(-U_0) - \theta_I(U_0) > 10^0$, then the dissipation associated with the motion of the contact line dominates that associated with the viscosity when $\frac{L}{a}\frac{U_0\mu}{\sigma} < 10^{-2}$.

5. SUMMARY

An approach has been presented by which the affect of the wettability on the displacement of immiscible fluids through a porous media can be assessed. It is limited to situations in which the Reynolds number is vanishingly small, and the capillary number is small. Under these conditions, the usual analytical difficulties associated with analyses of fluids spreading on solid surfaces are avoided.

Wettability refers to the mutual interactions between the two immiscible fluids and the solid. Its influences on the movements of the fluids occurs through the behavior of the dynamic contact angle. Thus, in so far as the contact angle plays an important role, wettability is an important issue.

Some implications of the characteristics of the functional dependence of the dynamic contact angle on the speed of the contact line were presented. It was shown that contact angle hysteresis is responsible for "globs" of fluid to appear to stick to the solid. It was also shown that contact angle hysteresis and the variation of the contact angle with contact line speed are both responsible for the dissipation of mechanical energy. Under conditions of very small capillary number, this dissipation dominates over that associated with the viscosities of the fluids.

REFERENCES

1. E. B. Dussan V., Ann. Rev. Fluid Mech. 11, 371 (1979).

2. P. G. de Gennes. Rev. Mod. Phys. 57, 827 (1985).

3. E. B. Dussan V. and S. H. Davis, J. Fluid Mech. 65, 71 (1974).

4. C. G. Ngan and E. B. Dussan V., submitted for publication.

5. C. G. Ngan and E. B. Dussan V., Phys. Fluids 27, 2785 (1984).

6. E. D. Murphy, Thesis, University of Pennsylvania, 1984 (unpublished).

7. E. B. Dussan V., J. Fluid Mech. 151, 1 (1985).

8. E. B. Dussan V., J. Fluid Mech. 174, 381 (1987).

9. E. B. Dussan V., and S. H. Davis, J. Fluid Mech. 172, (1986).

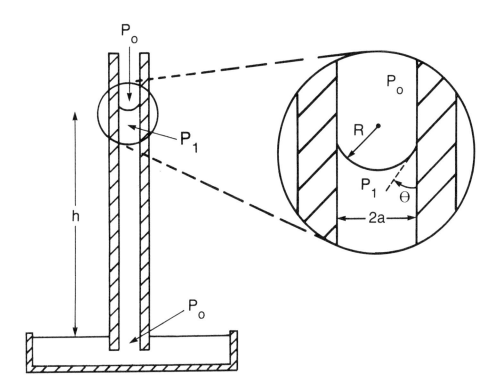

Figure 1. The lower end of a capillary of radius a is submerges into a bath of liquid causing the liquid to rise up the capillary to a height h. The pressures in the air and in the liquid just beneath the spherical cap shaped interface of radius R are P_0 and P_1, respectively. The contact angle, measured within the liquid, is θ.

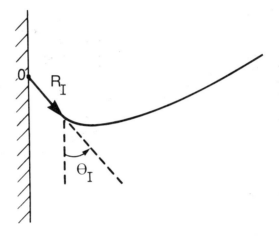

Figure 2. The point on the surface of the solid denoted by 0 represents the perceived location of the contact line. The boundary conditions for the shape of the interface in the outer region are that it passes through the point (R_I, θ_I) with an angle θ_I.

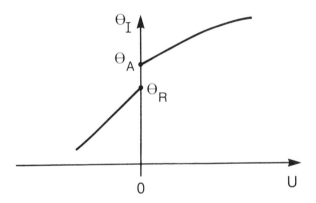

Figure 3. The assumed form of the dependence of θ_I on the speed of the contact line, U.

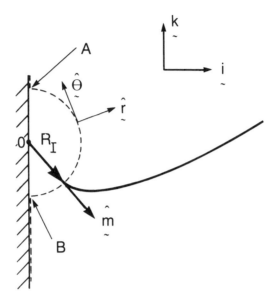

Figure 4. The calculation of the drag force exerted by the fluids on the surface of the solid requires the evaluation of $\int_{WALL} T_{zz} da$. This is accomplished by using the equation of conservation of linear momentum in integral form

$$-\int_{\substack{WALL \\ AB}} t_{zz}\, da + \int_{\substack{SEMI-CIRCLE \\ AB}} \underline{k} \cdot \underline{T}(\hat{r})\, da - \int_{\substack{line \\ AT(R_I, \vartheta_I)}} \sigma \underline{k} \cdot \hat{\underline{m}}\, d\ell = 0,$$

where the body forces and the rate of change of the linear momentum within a distance R_I of the perceived location of the contact line have been neglected; and $\hat{\underline{m}}$ denotes the unit vector pointing into the fluid body, tangent to the fluid interface, and perpendicular to the contact line.

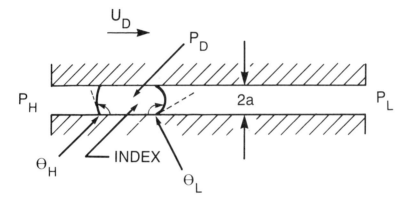

Figure 5. Two reservoirs are connected by a circular capillary of length L and radius a. If the difference in pressure between the two reservoirs, $P_H - P_L$, exceeds a critical value, then the index of fluid within the capillary moves at a speed U_D. The contact angles are measured from within the index.

STATISTICAL PHYSICS AND IMMISCIBLE DISPLACEMENTS THROUGH POROUS MEDIA

Roland LENORMAND
Dowell Schlumberger,
B.P. 90, 42003 Saint Etienne Cedex 1, France

ABSTRACT

A review is given of various mechanisms which occur during immiscible displacements in 2-dimensional networks of interconnected capillaries. We show how the physical laws governing meniscus equilibrium and flow conditions lead to different statistical models (mean field, percolation, D.L.A....).

INTRODUCTION

Many papers have been devoted to the use of Invasion Percolation [1-3] (I.P) and Diffusion Limited Aggregation [4,5] (D.L.A.) to describe the displacement of one fluid by another in a porous medium. These approaches are based on the randomness of the pore diameters of the porous solid associated with threshold effects for capillary forces (I.P) or viscous instabilities (D.L.A.).

In this paper I present some displacements in 2-dimensional porous media made of interconnected channels etched in a transparent plate (*micromodels*). First, I show how the various physical parameters describing the fluids and the geometry of the medium can explain the different mechanisms observed at microscopic scale. Second, I explain briefly how each mechanism can be described by a given statistical theory, not only the above-mentioned I.P. and D.L.A but also other forms of Percolation, Compact Cluster Growth, Mean Field Approach, etc. The main results are collected in table 1. Furthermore, the experimental results show the transitions between two mechanisms. For instance, by tuning only one parameter such as the flow rate, the pattern of the injected fluid evolves continuously from I.P. to D.L.A. (fig. 1). Finally, I show how the boundaries of the transition zones can be calculated in the case of drainage.

MONOPHASIC FLOW

The micromodels used in the experiments are made of transparent resin cast on a photographically etched mold [6]. The cross-section of each duct of the etched network is rectangular (fig. 2) with a constant depth ($x = 1$ mm) and a width d which varies from pore to pore with a given distribution and a random location. Various sizes of network are used for the experiments, the largest one containing more than 250,000 ducts. We will call *ducts* the cylindrical capillaries (bonds) and *pores* the volumes of the intersections (sites).

Table 1. Various displacement mechanisms and related statistical models.

	Type of Flow	Large Pores	Small Pores
MONOPHASIC FLOW	Flow in the ducts	EFFECTIVE MEDIUM APPROACH	
DRAINAGE	Flow in the ducts	INVASION PERCOLATION	
IMBIBITION	No flow by film	INVASION PERCO. Dual network	COMPACT GROWTH
	Flow by film	BOND PERCOLATION	COMPACT CLUSTER GROWTH
VISCOUS FLOW	Stable	FLAT INTERFACE	
	Unstable	GRADIENT GOVERNED GROWTH or DIFFUSION LIMITED AGGREGATION	

Porosity.

The simplest property of a porous medium is the *porosity* ϵ, the ratio between the volumes of the voids and the bulk. Plotting ϵ as a function of the radius ℓ of a disc centered on a point M chosen at a random location leads to the curves shown in fig. 3. For values of ℓ smaller than ℓ_0, of the order of the grain size a, one can see the microscopic effects due to the pore randomness. For values larger than ℓ_0 the porosity becomes constant and the medium can be studied as a *continuum*.

Permeability.

Another property which characterizes the porous matrix is the *permeability* k. When one fluid is flowing through a porous sample, the pressure drop ΔP between the two ends is linked to the length L, the cross-section S, the volumetric flow rate q and the viscosity μ of the fluid by Darcy's law: $\Delta P = \mu q L / k S$. From the geometrical properties, we can calculate the hydraulic conductance g

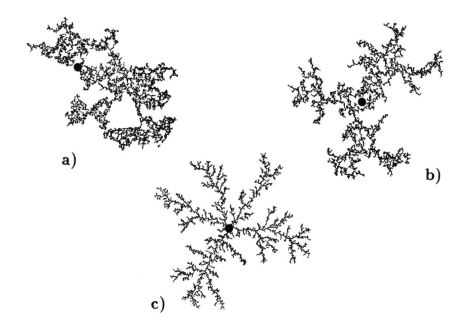

Fig. 1. Air displacing very viscous oil in a radial micromodel containing 250,000 ducts. a) very low flow rate: viscous forces are negligible and the pattern corresponds to invasion percolation; b) intermediate flow rate: crossover between percolation and D.L.A.; c) high flow rate: the pattern is very similar to D.L.A.

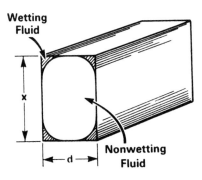

Fig. 2. Two fluid flow in a duct.

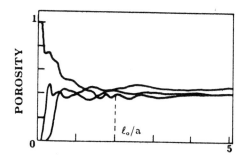

Fig. 3. Mean porosity inside a disc of radius ℓ in the micromodel as a function of the dimensionless radius ℓ/a, a being the mesh size of the square network.

of each channel of the network (analogous to a network of random resistors). The problem is to calculate the overall permeability k from the distribution law $f(g)$. This can be done by using an *effective medium* technique [7]. The principle is to look at only one channel (g) and replace all the other conductances by a mean value g_m. The local fluctuation due to the difference $g - g_m$ is calculated and we choose the value of g_m which leads to a zero average fluctuation. For the square network, this method leads to:

$$\int \frac{f(g)\ (g_m - g)\ dg}{g + g_m} = 0 \qquad (1)$$

CAPILLARY DISPLACEMENTS

In this case, viscous and gravity forces are assumed to be negligible and consequently all the displacement mechanisms are linked to capillary forces and randomness due to the different sizes of pores in a porous medium.

Generally speaking, when one fluid (say oil) is slowly displacing another immiscible fluid (say water) in a capillary tube of diameter D_0, the fluid for which the contact angle θ (between the tube and the meniscus) is smaller than $\pi/2$ is called the wetting fluid (W); the other one is the nonwetting fluid (NW). The pressure in the NW fluids exceeds the pressure in the wetting fluid by a value P, called capillary pressure and linked to the interfacial tension γ by the Laplace law: $P = 4\ \gamma\ \cos\theta/D_o$.

A displacement where the NW fluid is pushing the wetting fluid is called *drainage* (D), the reverse is *imbibition* (I). Figure 4 shows some close-ups of displacements.

Physical Mechanisms

Experiments show that both fluids can flow simultaneously in the same duct with different velocities, the wetting fluid remaining in the extreme corners of the cross-section and roughness of the walls (fig. 2). This effect explains the richness of mechanisms involved during displacements.

A displacement can be divided into three parts: i) flow of the injected fluid from the entrance towards the moving meniscus, ii) displacement of the meniscus, iii) flow of the displaced fluid towards the exit.

Meniscus Displacements.

This study have been presented in detail in previous papers [8,9] and the main result is the effect of the geometry of the pores which can introduce a selection of imbibition mechanisms. We found two main cases: i) *large pores*, when the size of the pore is large compared with duct diameters; the meniscus is unstable inside a pore (fig. 5a). ii) *small pores*, in the opposite case, the meniscus collapses inside a duct (fig. 5b). We will see later the consequences of these two

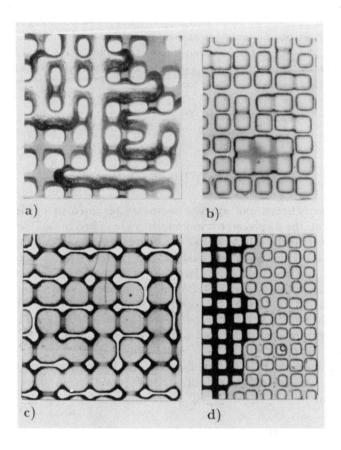

Fig. 4. Close-ups of various experiments in micromodels with air (in white) and oil (in grey). a) drainage; b) imbibition at very low flow rate in "small pore" geometry: the wetting fluid (oil) flows as a film along the walls and collapses in the smallest ducts. Clusters are either single ducts or rectangles (compact cluster growth). c) imbibition at very low flow rate in "large pores" geometry: the W fluid (grey) collapses in smallest ducts without filling pores (classical percolation). d) imbibition without flow by film: the succession of collapses insites pores leads to a flat interface at large scale.

cases, the former leading to percolation type patterns, the later to flat interfaces at large scales (table 1).

Flow of the fluids.

The nonwetting fluid can flow only in the bulk of the ducts. So the flow occurs only if a continuous path of ducts or pores filled with this phase exists either towards the entrance (during drainage) or the exit (during imbibition) of the network. Otherwise the NW fluid is trapped.

We observe 3 kinds of flow of the wetting fluid: i) flow in the bulk of the ducts, ii) along the corners when the NW fluid fills the central part of the duct, iii) by *film* along the roughness of the walls, but only in case of strong wettability and very low flow rate.

Drainage

Capillary forces prevent the NW fluid from spontaneously entering a porous medium. It can only enter a duct (diameter D_o) when the pressure exceeds the capillary pressure. From a statistical point of view a duct with $D > D_o$ is an active or conductive bond and a duct with $D < D_o$ an inactive bond. The fraction p of active bonds can easily be deduced from the throat size distribution.

At a given pressure P, the injected fluid invades all the percolation clusters connected to the injection point (fig. 1a at large scale and fig 4a at pore scale); this mechanism has been called invasion percolation [1-3,10]. During the displacement, the wetting phase is trapped in the network when the invading NW fluid breaks the continuous path toward the exit (in 2-D, the W fluid cannot escape by flowing via the corners).

Imbibition

The type of displacement depends upon two factors: the pore geometry (large or small pores) and the possibility of flow by film along the roughness. Simulations of the 4 types of displacement in small networks are shown in fig. 6.

No flow by film.

For *large pores* (fig. 6a), the injected wetting fluid invades the network by a succession of collapses in the ducts (fig. 5b). Due to pressure effects, the smallest channel of the interface between the fluids is filled at each step and because of the randomness of the pore size in the network, this mechanism can be described by *invasion percolation*. However, the topology is not the same as in drainage, and we have shown [9,11] that the process corresponds to *bond percolation* in the *dual* network (the sites of the dual network are the centers of the solid grains).

For small pores (fig. 6b), the meniscus instability inside a pore (fig. 5a) leads to filling the network line after line, without trapping. The result is a kind of

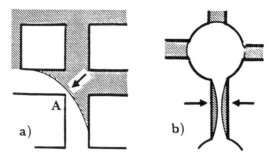

Fig. 5. Different mechanisms for the meniscus displacement during imbibition: a) small pore geometry: the meniscus collapses inside the intersection (pore) when it reaches point A of the walls; b) large pore geometry: the meniscus collapses inside a duct.

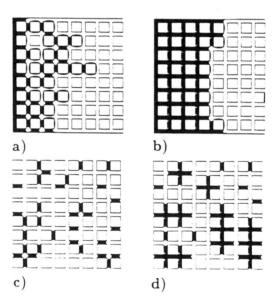

Fig. 6. Simulation of the different imbibition mechanisms. The injected wetting fluid is black. a) no flow by film, large pores (*invasion percolation*; b) no flow by film, small pores (*flat interface*); c) flow by film, large pores (*classical percolation*); c) flow by film, small pores (*compact cluster growth*).

compact crystal growth, with a shape related to the network mesh (fig. 4d).

Flow by film.

In this case, the wetting fluid can always flow in the bulk of the ducts or along the corners but it can also spread along the roughness of the walls and reach all the ducts in the network without continuity problems. The thickness of the film increases with time until a collapse occurs in the smallest duct and so forth. The resulting pattern depends on meniscus stability in the pore, related to the geometry:

For *large pores*, the meniscus remains stable in the pores, even if many ducts are filled with the wetting fluid (experiment fig. 4c and simulation fig. 6c). The number of filled ducts increases with time at random locations until the NW fluid becomes disconnected. The result is a *classical percolation* process.

For *small pores*, the meniscus is unstable inside a pore when two adjacent ducts are filled and very quickly, the pore and the two other ducts are filled (fig. 6d). This leads to rectangle clusters which grow with time (experiments fig. 4b). For a critical fraction p^* of collapsed ducts, one of the clusters "swallows" the other and invades the entire network. We have shown [12] that p^* does not have a finite value but tends to zero as the network size N increases ($p^* \propto 1/logN$). Simulation, fig. 7, shows the situation just before the threshold. We called this mechanism *compact cluster growth*.

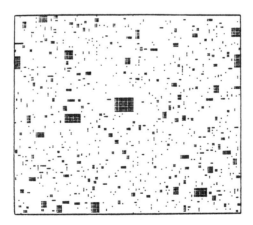

Fig. 7. Simulation of the compact cluster growth just below the threshold in a square network (150x150). The black dots represents the filled pores (sites).

At a given stage of a capillary displacement, menisci in the pores do not "see" the exit because we are assuming a zero pressure drop in the fluids. Consequently, they are described by "local models", i.e, at each step, the interface between the two phases moves towards an adjacent site according to some local rules. The mechanism is different for viscous displacements.

VISCOUS DISPLACEMENTS

In this case, capillary forces are negligible compared to viscous forces. However, we need a small amount of capillary effects, otherwise the fluids mix inside the ducts and the problem is more complex (miscible displacements).

Viscous displacements, either stable or unstable are governed by the pressure field between the entrance and the exit. Consequently, even in the case of a stable displacement, a local model based on some rules at the interface cannot be used for modeling viscous displacements. A model, called *Gradient Governed Growth* has been developed simultaneously by several authors [13-15] to solve this problem, using both a continuum approach to calculate the pressure field and a discrete displacement of the interface which accounts for the granular structure of the porous medium.

Injecting the more viscous fluid.

The displacement is stable and leads to a *flat interface* at large scale. However, we expect fluctuations due to permeability variations at the microscopic scale ℓ_0 previously defined (porosity).

Injecting the less viscous fluid.

The displacement is unstable (analogy with Saffman-Taylor [16] instability in a Hele Shaw cell). If we assume that the injected fluid has negligible viscosity, the growing pattern can also be represented by a model known as *Diffusion Limited Aggregation* [4,17]. A computation [19] based on a network of *random* conductances, leads to similar results. The validity of this model is demonstrated by the similarity between computer simulations and experimental patterns. For instance, fig 1c shows the displacement of a very viscous oil by air in a radial geometry.

Now, in the case of drainage, let us calculate the boundaries of the different domains.

THE PHASE DIAGRAM

So far we have identified three main regimes for the displacement of a wetting fluid by a non-wetting fluid (drainage) and presented the corresponding models:

- stable displacement and G.G.G without randomness (continuum approach).

- viscous fingering and D.L.A. (or $M \to 0$ limit for G.G.G.).
- capillary fingering and invasion percolation.

These three basic mechanisms can be understood as theoretical limits when two of the three kinds of forces involved during the displacement are negligible and theoretically, these limits cannot be reached in an infinite medium. However, as shown in the 2-D experiments [6], these basic mechanisms can be observed in a finite system. The purpose of the following section is to calculate the frontiers of these domains to answer the important question: for given conditions of injection, which kind of mechanism should we expect?

As previously described, each *pure* or basic mechanism is characteristic of a dominant type of forces: viscous effect in the displacing fluid for plug flow, viscous forces in the displaced fluid for D.L.A and capillary forces for percolation. However, during an experiment, the three kinds of forces are always acting together and writing a general condition such as $Ca \ll 1$ or $M \ll 1$ does not provide useful information and, furthermore, can be dangerous in some cases. For instance, experiments in real porous samples have shown the influence of viscous forces when $Ca < 10^{-6}$ (and obviously $Ca \ll 1$!).

Previously, in a short note [18], we calculated approximately the boundaries between these three different mechanisms by using only two parameters, the capillary number Ca:

$$Ca = \frac{q\,\mu_2}{\Sigma\,\gamma \cos\theta} \qquad (2)$$

where Σ is the cross-section area of the sample q/Σ being the mean velocity of the fluid in the channels, and $M = \mu_2/\mu_1$ the viscosity ratio between the two fluids. This approach led to an original display of the domains on a general diagram with axes representing Ca and M and we called this result the "phase-diagram" for capillary and viscous displacements in 2-dimensional porous media.

In this study, we propose a 3-dimensional version of the "phase-diagram" [20]. The model is based on the calculation of the various forces during a displacement in a 3-dimensional porous medium, assuming a simple form for the fingers. Using very naive assumptions we estimate the boundaries of the various domains as functions of the fluids properties (viscosity ratio M, capillary number Ca), the sample size L and parameters related to the pore geometry (permeability, aspect ratio...).

The frontiers are strongly linked to the geometrical properties of the medium which will be presented in the first part of this section. Then, the limits of the three domains will be calculated in turn.

Properties of the Medium

The porous medium is described by a network of interconnected capillaries with diameters D varying with a distribution law $f(D)$ which represents the probability of finding a channel with a diameter between D and $D + dD$ (fig. 8):

$$\int f(D)\,dD = 1 \qquad (3)$$

The macroscopic size of the network is $L \times L \times L$ and the mesh size is a (microscopic length scale). For simplicity reasons, we assume that the maximum value of the pore diameter is close to the mean value D_0. The mean width ΔD of the pore size distribution is defined by:

$$f(D_0) \, \Delta D = 1 \tag{4}$$

For a monophasic flow, the pressure gradient $\Delta P/L$ across the network is linked to the volumetric flow rate q, the permeability k, the cross-section area of the flow Σ and the viscosity μ of the fluid by Darcy's law:

$$\frac{\Delta P}{L} = \frac{\mu}{k} \frac{q}{\Sigma} \tag{5}$$

Percolation Domain

Let us consider the quasi-static limit of the displacement ($Ca \to 0$), as it could be simulated by computer. Let us call n the number of displaced channels at breakthrough. As a consequence of the *fractal*[21] nature of the percolation cluster, the fraction n/N tends to zero when the size of the network goes to infinity. Introducing now a small amount of viscous effects (by increasing the flow rate) changes the value of n by a small amount δn:

- Injecting the more viscous fluid ($M \to \infty$).

  Close to the percolation threshold, the injected fluid flows through the "backbone" of the percolation cluster, and, due to viscous effects, the pressure is larger near the entrance. Consequently, some channels, with a size smaller than the threshold value D_c, can be now invaded and the fraction of accessible channels increases by a value δp. The pattern is more compact near the entrance compared to the pure percolation case. Increasing the flow rate will progressively lead toward a stable displacement (cross-over between percolation and plug flow).

- Injecting the less viscous fluid ($M \to 0$).

  The capillary pressure is decreased by δP due to the pressure drop in the *displaced* fluid. Consequently, the fraction of accessible channels decreases by a value δp and the fingers become thinner and thinner when the flow rate increases (towards the D.L.A. domain).

In both cases, increasing the flow rate changes the fraction p of accessible channels in the neighborhood of the percolation threshold p_c and consequently leads to a change in the number n of displaced channels. If this variation is very small (say for instance $\delta n/n = 10^{-6}$), we cannot discern the difference between the true percolation pattern and the real pattern neither by visual examination nor by a macroscopic measurement such as the volume of displaced fluid. From a theoretical viewpoint, we have to fix an arbitrary threshold, for the ratio $\delta n/n$ when we we are just capable to detect a difference either by visual examination or macroscopic measurement. This threshold will correspond to the boundaries of the percolation domain.

The relationship between δp and δn must take into account the critical behavior near the percolation threshold (critical exponents). For simplicity reasons we do not present here the details of the calculation. We only assume that the boundaries of the percolation domain correspond to a given limit for the variation δp of accessible channels. We will first estimate the value of δp in a general case, and then, we will obtain the equations of the frontiers $Ca^* = f(M)$ by writing the condition $\delta p = \delta p^*$.

Injecting the more viscous fluid $(M \to \infty)$.

The viscosity of the injected fluid (2) is much larger than the displaced fluid (1), so we neglect the viscous pressure drop due to the displaced fluid. The calculation is based on several naive assumptions: i) Close to the percolation threshold, the injected fluid flows through the "backbone" of the percolation cluster which is replaced by a straight channel with a length L. ii) All the flow rate q occurs through this channel. iii) The pressure drop δP between the ends of the channel is calculated by using Darcy's law (equ. 5) for the flow through a cylinder of length L and effective cross-section area $\Sigma = a^2$:

$$\delta P = \frac{L \mu_2 q}{k a^2} \qquad (6)$$

The increase of pressure in the injected fluid leads to an increase of the capillary pressure $\delta P_c = \delta P$ and, consequently, the size of the accessible channels decreases by a quantity δD, calculated from Laplace's equation:

$$\delta P_c = -\frac{2 \gamma \cos\theta}{D_c^2} \delta D \qquad (7)$$

The variation of accessible channels is given by the area of the hatched zone in figure 8:

$$\delta p = -\delta D \, f(D_c) \qquad (8)$$

and combining the previous equations:

$$\delta p = \frac{D_c^2 \, a \, f(d_c)}{2 \, k} \times \frac{L^3}{a^3} \times \frac{\mu_2 \, q}{L^2 \, \gamma\cos\theta} \qquad (9)$$

In this equation, the last factor is the capillary number (equ. 2) and the first one is only a function of the geometrical properties of the porous matrix. Let us define the geometrical constant G:

$$G = \frac{2 \, k}{D_0 \, a} \times \frac{\Delta D}{D_0} \qquad (10)$$

and, using equ. 4:

$$G' = \frac{D_0^2}{D_c^2} \times \frac{f(D_0)}{f(D_c)} \qquad (11)$$

The second constant G' contains percolation properties (especially the connectivity of the network).

Now we assume that $\delta p = \delta p^*$ on the boundary of the percolation domain, which leads to the capillary number labeled Ca^* on the frontier of the percolation domain towards plug flow:

$$Ca^* = \delta p^* \, G \, G' \, \frac{a^3}{L^3} \qquad (12)$$

Injecting the less viscous fluid ($M \to 0$).

The calculation is almost the same as previously, except that the capillary pressure is decreased by δP due to the pressure drop in the *displaced* fluid (viscous effects are neglected in the injected fluid).

The pressure drop is calculated by assuming that the displaced fluid can flow in the whole network, without reduction of permeability due to the branches formed by the injected fluid (the displaced fluid is very far from the percolation threshold). From equ. 5:

$$\delta P = \frac{\mu_1 \, q \, L}{k \, L^2} \qquad (13)$$

Now, using the same method as previously, with $\mu_1 = \mu_2/M$, leads to the equation of the boundary of percolation towards D.L.A:

$$Ca^* = \delta p^* \, G \, G' \, M \, \frac{a}{L} \qquad (14)$$

D.L.A. Domain

The analogy between viscous fingering and D.L.A simulations requires the following three conditions: 1) No pressure drop in the injected fluid (limit $M \to 0$). 2) No pressure jump at the interface between the fluids (Capillary forces $\ll$ viscous forces). 3) Large randomness due to the pore size distribution [18].

Conditions 1 and 2 determine the two frontiers of the D.L.A domain, the viscous and the capillary limits. Condition 3 is assumed to be always satisfied in a porous medium.

Condition 1–Viscous Limit.

The pressure drop in the displacing fluid must be much smaller than the pressure drop in the displaced fluid. The capillary forces are assumed to be negligible.

The injected fluid flows both at the tip of the tree-like finger and everywhere on the interface between the fluids with respective flow rates q' and q''. In a pure D.L.A, only the tip of the finger is growing (screening effect) and consequently $q'' = 0$. So, the effect of the finite viscosity ratio can be parameterized by the ratio q''/q'. This ratio is estimated by using the approximations of two independent flows, one in the finger (replaced by a straight channel) and the other in the whole network. The pressures are calculated by using equ. 6 and

13 with respective flow rates q' and q'' and respective viscosities μ_2 and μ_1. The equality of pressure at the entrance leads to:

$$\frac{q''}{q'} = M \frac{L^2}{a^2} \qquad (15)$$

Now, the boundary of the domain is obtained by fixing an arbitrary threshold ϵ for the ratio q'/q''. The value of ϵ is linked to the accuracy of the observation, for instance the measurement of the volume of displaced fluid. With this condition, we obtain the equation for the frontier between D.L.A and plug flow domains:

$$M^* = \epsilon \frac{a^2}{L^2} \qquad (16)$$

Condition 2–Capillary Limit.

Condition 2 requires no pressure jump at the interface between the fluids. To be more accurate, it seems that a constant and small pressure jump would not perturb a D.L.A. type displacement (the interface is always an equipotential). Consequently, fluctuations from pore to pore must be negligible compared to the viscous pressure. Let us evaluate the mean fluctuation δP_c of the capillary pressure.

Because the viscous pressure drop is proportional to D^{-4}, the injected fluid enters only the largest channels and finds a continuous path, so we can assume $D \geq D_0$. Now, as illustrated in fig. 9, a perturbation can occur in some very large channels. Without capillary effects, the injected fluid would enter the pore labeled 1 if the equipotential were locally perpendicular to the mean flow direction, assuming that the displacement takes place in only one channel at each time step. However, with capillary forces, the fluid would prefer to enter the largest channel (labeled 2) if:

$$P_{c1} - P_{c2} < \delta P \qquad (17)$$

To obtain a visible effect at macroscopic scale, the amount of large pores 2 ($D > D_L$) in the network must be at least a few %. So, D_L can be estimated by the relation $D_L - D_0 \approx \Delta D$, (calculated with a Gaussian pore size distribution, ΔD being equal to twice the standard deviation). Consequently, the mean fluctuation due to capillary effects, is given approximately by:

$$\Delta P_c = 2 \gamma \cos\theta \frac{\Delta D}{D_0^2} \qquad (18)$$

and must be smaller or at least equal (on the frontier) to the viscous pressure drop δP at the pore scale in the displaced fluid (from equ. 5):

$$\frac{\delta P}{a} = \frac{\mu_1 \, q}{k \, a^2} \qquad (19)$$

This condition lead to the equation of the frontier:

$$2 \gamma \cos\theta \frac{\Delta D}{D_0^2} = \frac{\mu_1 \, q \, a}{k \, a^2} \qquad (20)$$

$$Ca^* = M\,G\,\frac{a^2}{L^2} \qquad (21)$$

Stable Domain

In this domain, the dominant force is linked to the viscosity of the injected fluid and perturbation are provided by capillary effects or viscous effects in the displaced fluid.

Capillary limit.
The injected fluid invades all the channels in the network. Therefore, capillary effects are due to the difference of capillary pressure between the smallest and the largest channels. As before, we assume that this effect can be observed when the relative amount of concerned pores is larger than a few %. For a Gaussian pore size distribution, this approximately leads to:

$$\Delta P_c = 4\,\gamma\,\cos\theta\,\frac{\Delta D}{D_0^2} \qquad (22)$$

On the frontier, ΔP_c is of the order of the pressure drop at the pore scale (we assume $\ell \approx a$). However, contrary to the fingering case, we observe a displacement in *all the pores* on the frontier at the same time (flow rate $\approx q\,a^2/\Sigma$ in each pore). Therefore, the pressure drop on a length scale equal to one pore is given by:

$$\delta P = a\,\frac{\mu_2}{k\,a^2}\,\left(\frac{q\,a^2}{L^2}\right) \qquad (23)$$

The frontier is obtained by combining the two previous equations:

$$Ca^* = 2\,G \qquad (24)$$

Viscous Limit.
In this limit, capillary forces are assumed to be nil. However, this case is not a "miscible" displacement and the fluids cannot mix. Consequently, there is no characteristic time scale linked to molecular diffusion and, for a creeping flow (no inertial forces), the displacement pattern is independent of the flow rate (vertical frontier).

Further simulations with a G.G.G. model must be performed for a better understanding of this viscous limit. However, the classical value $M = 1$ for the onset of instabilities in a continuum approach seems an acceptable result.

The Phase Diagram

The frontiers of the three domains are displayed in the plane with axes Ca and M in log-log scale (fig. 10). The bold lines represents the calculated limits and the dashed lines are estimates of the intermediate zones. The diagram displays the three domains where the mechanisms can be considered as pure:

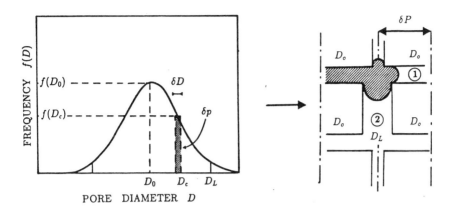

Fig. 8. Pore size distribution.

Fig. 9. Displacement mechanism during plug flow.

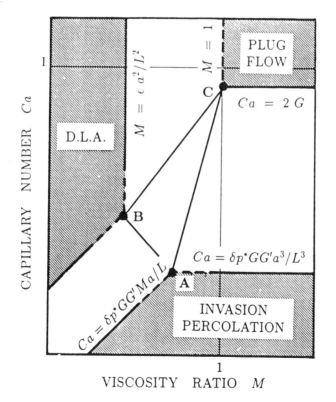

Fig. 10. Phase diagram for drainage.

i) plug flow for favorable viscosity ratio and high capillary number, ii) percolation for low Ca, iii) D.L.A for low viscosity ratio and high Ca.

Each domain is limited by two frontiers and the coordinates of the intersections seems to be of great importance for scaling problems:

- Percolation domain: point **A**.
  $Ca = \delta p^* \, G \, G' \, a^3/L^3 \, ; \, M = a^2/L^2$

- D.L.A domain: point **B**.
  $Ca = \epsilon \, G \, a^4/L^4 \, ; \, M = \epsilon \, a^2/L^2$

- Plug flow domain: point **C**.
  $Ca = 2 \, G \, ; \, M = 1$

An experiment or a real displacement is represented by a point on this diagram and its location gives information on the physical mechanisms involved during the displacement:

- inside a domain, only one type of force is involved, capillary forces for percolation, viscous forces in the injected fluid for plug flow and viscous forces in the displacing fluid for D.L.A.

- outside the domains and outside the triangle ABC, two types of forces are involved (transition zones).

- inside the triangle ABC, the three types of forces are involved.

CONCLUSION

We have presented a theoretical description of immiscible displacements (drainage) in 3-D porous media taking into account viscous and capillary forces. We have shown how the recent stochastic theories (especially percolation and diffusion limited aggregation) can be included in a general description of displacement patterns. This approach takes into account the fluids and flow properties and also the pore geometry.

However, this study needs further investigations; the determination of unknown parameters (such as δp^* related to the percolation threshold) and also experimental verifications in real 3-dimensional porous media.

ACKNOWLEDGMENTS

This work has benefited from many conversations with Cesar Zarcone and colleagues at Schlumberger Doll Research (Ridgefield) and Dowell Schlumberger (Saint Etienne). We are especially indebted to Yves Pomeau for many discussions which led to the notion of "phase diagram". This research has been partially supported by Centre National de la Recherche Scientifique (France).

REFERENCES

1. R. Lenormand and S. Bories, C. R. Acad. Sc. Paris **291B**, 279 (1980).
2. D. Wilkinson and J. F. Willemsen, J. Phys. A **16**, 3365 (1983).
3. J. T. Chayes, L. Chayes and C. M. Newman, Commun. Math. Phys. **101**, 383 (1985).
4. L. Paterson, Phys. Rev. Lett. **52**, 1621 (1984).
5. K. J. Måløy, J. Feder and T. Jøssang, Phys. Rev. Lett. **55**, 1885 (1985).
6. R. Lenormand and Zarcone, Proc. P.C.H. conf., Tel-Aviv, (1984), Phys. Chem. Hydrodyn. **6**, 497 (1985).
7. S. Kirkpatrick, Rev. Modern Phys. **45**, 574 (1973).
8. R. Lenormand, C. Zarcone and A. Sarr, J. Fluid Mech. **135**, 337 (1983).
9. R. Lenormand and C. Zarcone, Soc. Petrol. Eng. paper 13264 (1984).
10. R. Lenormand and C. Zarcone, Phys. Rev. Let. **54**, 2226 (1985).
11. R. Lenormand, C. R. Acad. Sc. Paris **291B**, 279 (1980).
12. R. Lenormand and C. Zarcone in Kinetics of Aggregation and Gelation (F. Family and D.P. Landau ed., Elsevier, 1984).
13. A. J. DeGregoria, Phys. Fluids **28**, 2933 (1985).
14. M. King and H. Sher, Soc. Petrol. Eng. paper 14366 (1985).
15. J. D. Sherwood and J. Nittmann, J. Physique **47**, 15 (1986).
16. P. G. Saffman and G. I. Taylor, proc. R. Soc. Lond. **A245**, 311 (1958).
17. T. A. Witten and Sander, Phys. Rev. B **27**, 5686 (1983).
18. J. D. Chen and D. Wilkinson, Phys. Rev. Lett. **55**, 1892 (1985).
19. R. Lenormand, C. R. Acad. Sci. Paris série II, **301**, 247 (1985).
20. R. Lenormand, Soc. Petrol. Eng., paper 15390 (1985).
21. B. B. Mandelbrot, The Fractal Geometry of Nature (Freeman, San Francisco, 1982).

STABLE AND UNSTABLE MISCIBLE FLOWS THROUGH POROUS MEDIA [*]

J.-C. Bacri, N. Rakotomalala, D. Salin

Laboratoire d'Ultrasons, Université Pierre et Marie Curie,
Tour 13, 75252 Paris Cedex 05, France

INTRODUCTION

Displacements of miscible fluids through porous media concern many industrial channels such as secondary oil recovery, chemical engineering and dispersion of either chemical or nuclear polluant (very up-to-date !). The basic problem is to find out in which way a fluid, present in a porous medium, is displaced by an injected fluid miscible to the first one :
i) In the limit of zero flow rates, the mixing of the two fluids is due to molecular diffusion, affected by the tortuosity of the porous medium.
ii) When the injected fluid is more viscous than the displaced one, the displacement is stable and the mixing of the two fluids is enhanced by convection of the fluids along the many different flow passages through the porous medium. This mechanical mixing refers to hydrodynamical dispersion.
iii) When the injected fluid is less viscous than the displaced one the displacement can be unstable : viscous fingers develop into the porous medium.

Pioneer works of Taylor[1] and Saffman[2] have accounted for the dispersion in pipe and in a model porous medium (tubes at random directions). The same authors[3] and others[4] also explain the main features of the viscous fingering in the Hele-Shaw experiment. In the early eighties the understanding of both dispersion and fingering profits by percolation concept and fractal fashion : de Gennes[5] and Sahimi et al[6] have analyzed the dispersion in a porous medium close to a percolation threshold ; the dispersion is the result of flow mixing on a so-called backbone and of molecular diffusion into the dead-ends ; these notions are very fruitful for the interpretation of dispersion in unsaturated porous media. Paterson[7] has described the unstable two fluid displacements in a porous medium, using the Witten and Sander idea[8] of random walker in the so-called diffusion limited aggregation (DLA). These two concepts gave new vitality to both subjects (dispersion and instability) which will be most likely demonstrated in this symposium. From the experimental point of view dispersion has been studied essentially in unconsolidated media[9-12] and most of the experiments are carried out through effluent

---

[*] Work partially supported by Dowell-Schlumberger, Z.I. de Molina la Chazotte, BP 90, 42003 Saint-Etienne Cedex, France.
[†] Associated with the Centre National de la Recherche Scientifique.

techniques[9,13] which miss the spatial aspect of the dispersion. Unstable displacements have been carefully observed and described in two dimensional etched network[14] and different Hele-Shaw like cells[7,15] ; then this 2D-problem has been understood with the help of numerical simulations[15,16] ; but in three dimensions...

In this paper, by means of an acoustical technique, which allows us[12,17] to determine the fluid concentrations all along the core of 3D-porous media with a 0.1 inch spatial resolution, we report experimental measurements on both stable and unstable miscible fluid flows in consolidated and unconsolidated porous media. Using a third immiscible fluid it has been possible to change the morphology of network on which dispersion occurs in order to test percolation theory predictions. Although this paper is devoted to show a large quantity of data in order to stimulate 3D-calculations, we will first summarize theoretical aspects, then describe our experimental set-up and procedures, give our experimental results and discuss them.

THEORETICAL SURVEY

1.- Hydrodynamic dispersion

Dispersion in a disordered, multiconnected, statistically homogeneous porous medium is classically described by a macroscopic convection-diffusion equation[1-3, 9-13, 18-19] :

$$\frac{\partial C}{\partial t} + U \frac{\partial C}{\partial x} = D_{\parallel} \frac{\partial^2 C}{\partial x^2} + D_{\perp} \nabla_{\perp}^2 C \qquad (1)$$

where C is the space and time dependent fluid concentration, U the average Darcy fluid velocity in the x-direction, $\nabla_{\perp}^2$ the Laplacian in transverse direction ; coefficients $D_{\parallel}$ and $D_{\perp}$ characterize longitudinal and transverse dispersions. Equation (1) accounts for normal, i.e. gaussian, dispersion : statistical approach is an appropriate description of the chaotic nature of the flow passages through the pores. The basic problem of dispersion is to determine $D_{\parallel}$ and $D_{\perp}$ whatever the flow rate U is. Two regimes have been fairly explored :

i) As U tends towards zero, molecular diffusion overcomes and $D_{\parallel}$ becomes constant :

$$D_{\parallel} = D_m / \alpha_a \qquad (2)$$

D_m is the molecular diffusion coefficient and α_a (> 1) is a reducing factor accounting for the tortuous nature of the porous medium. Molecules spend a lot of time to diffuse in the tortuous branches of the medium ; this statistical problem corresponds to the one of the random walk of an ant in a labyrinth[20-22] : this problem has been solved in the vicinity of percolation thresholds[21]. The tortuosity α involved in (2) has been related[23] to the formation factor F ($\alpha = F\phi$, ϕ porosity) of the medium ; F is the ratio of the conductivity of a brine to the conductivity of the (insulating) porous medium saturated with the same brine. This relationship is to our opinion questionable : the tortuosity α_a defined from diffusion processes involves the whole pore space, backbone and dead-ends in the percolation language ; conductivity measurements, as well as permeability, involve a kind of mean passage through the network and do not care of dead-ends ; then we conjecture that the formation factor is related to the tortuosity of the backbone α_{BB} ($\alpha_{BB} = F\phi$). We have to notice that from the experimental point of view quite a lot of measurements can lead to a tortuosity : acoustic and fourth sound velocity[24], conductivity and permeability, dielectric permittivity and diffusion at zero flow rate. These measurements have to be related to different theoretical tortuosities[20-22,25].

ii) At large flow rate U, mechanical mixing dominates : fluid particles are convected along different streamlines ; sometimes particles move faster than the average flow U, sometimes slower : mechanical dispersion is due to the different transit times along these flow passages. In case of packed beds[11], different phenomena contribute to dispersion such as stagnation points, close streamlines, boundary layers... Aware of these complicated contributions one can anyway give a rough but simple dependence of the longitudinal dispersion coefficient $D_{/\!/}$ on the flow velocity U :

$$D_{/\!/} = U \cdot \ell_D \qquad (3)$$

where the determination of the dispersion characteristic length, ℓ_D, is the crucial problem. In unconsolidated media, such as packed beads, theory[11] and experiment lead to $\ell_D \sim d$,[9-13] the typical diameter of the beads ; but in fact in such porous media neither the tortuosity ($\alpha \sim 1$) nor the disorder (roughly monodisperse grains) can influence on dispersion. In a bundle of straight tubes of constant cross-section, statistical approach of Saffman[2] accounts for the interplay between molecular diffusion and convection ; the typical length in (3) is the tube radius. In fact such models are far away from the disordered morphology of a consolidated porous medium which is more relevant to percolation approach. Recent calculations[26] on a network of tubes of distributed cross-sections show that the larger the width of the distribution, the larger the dispersion ; this is an insight of the importance of disorder on dispersion (small flow passages contribute widely to dispersion). Previous Monte-Carlo simulations of Sahimi et al[6] lead to permeability and dispersion in percolation networks. Moreover percolation concepts are very useful for experimentalists because using a third immiscible fluid we can change the morphology of the labyrinth on which dispersion will occur, without changing the porous medium. Depending on the immiscible fluid injection processes (imbibition or drainage) we can reach percolation threshold in fluid saturation ; in the vicinity of percolation threshold, dispersion coefficient is expected to diverge[5,6] in a way which depends drastically on the tortuosity of mazes of the percolating network : de Gennes[5] found a smooth divergence (exponent $(\bar{t} - \beta - 2\nu)/\beta \sim -0.5$ in 3D) ; reconsidering the problem of the backbone tortuosity, Vannimenus[27] expected a larger divergence (exponent -2) ; Sahimi et al gave a numerical simulation of this divergence.

2.- Unstable displacement

When the injected fluid is less viscous (μ_I) than the displaced one (μ_D) the interface of the two fluids can present a fingering instability (Saffman -Taylor[3-4]). Paterson[7] remarked that flow equations in porous media (Darcy law and uncompressibility) lead to the same Laplacian equation than the movement of random walker in diffusion limited aggregation[8] (DLA) : as a consequence, the structure of the interface - pattern is fractal with an exponent which depends on the dimensionality. Experiments on two dimension transparent etched network[14] have verified these conjectures. The Paterson-derivation supports that there is no interfacial tension between the two fluids (i.e. miscible) and that the less viscous fluid is inviscid ($M = \mu_D/\mu_I = \infty$) ; this last condition rules out any influence of flow rate on DLA . If the literature on DLA and fractal structure is superabundant, few papers have been devoted to the transition from DLA ($M = \infty$) to the stable ($M < 1$) displacement, piston-like in Lenormand's "phase diagram"[28]. De Gregoria[29] reported on sweep efficiency variations with M ; more useful for experiment comparison, Sherwood and Nittmann[16] give the profile of concentration along the sample for different M ratios ; unfortunately this is only a 2D-simulation.

EXPERIMENT

We have investigated both unconsolidated porous media (glass beads, broken glass) and consolidated ones (sandstones). Table 1 summarizes our measurements of porosity ϕ, permeability k, formation factor F, and the corresponding conductivity tortuosity deduced from $\alpha = F \cdot \phi$. The characteristic pore size ℓ_c of the medium has been derived recently by different authors[30, 31] from permeability k and conductivity measurements :

$$\ell_c^2 = a \cdot F \cdot k \qquad (4)$$

where a is a constant; a = 1/226 for a percolation-like broad distribution of pore radii. When applied to a random pack of spheres of

Porous medium	ϕ	k (Darcy)	F	α	ℓ_c (μm)	ℓ_D (μm)
200 μm glass beads	0.40	40	4	1.6	200	100
Vosgian sandstone	0.21	0.11	35	7	15	600

Table 1. *Measured characteristics of our porous media : porosity ϕ, permeability k (1 Darcy = 10^{-12} m^2), formation factor F (deduced from conductivity measurements), conductivity tortuosity $\alpha = F \phi$. ℓ_c is the characteristic size of the grains deduced from k and F through equation (4). ℓ_D is the dispersion length deduced from dispersion coefficients using equation (3).*

diameter d, equation (4) leads to $\ell_c \simeq 0.99$ d, exactly the sphere diameter. Sandstones have been used, either fully saturated with miscible fluids, or unsaturated by injection of a third immiscible fluid. Depending on the way of displacing the immiscible fluid in the sandstones (imbibition or drainage at different flow rates[17]) we obtain different immiscible fluid repartitions; the non-wetting phase (saturation S_0) behaves as a maze, the wetting phase (saturation $S_w = 1 - S_0$) behaves as the complementary maze. We study the dispersion on the non-wetting phase maze using a couple of non-wetting fluids.

Porous media samples are core samples of size $4 \times 4 \times 25$ cm^3; flows are driven vertically along the largest dimension at constant flow rates from 0.2 to 1200 cm^3/h. Saturation measurements are derived from the velocity variations of a sound wave in ten cross-sections of the sample. The acoustic wave is generated and analyzed thanks to ten transmitter-receiver pairs of transducers laid every 2.5 cm along the larger dimension of the sample. The spatial resolution in this direction reaches 3 mm; the ten positions are scanned in less than two minutes through automatic procedure. Calibration curve of the velocity variations with C and the accuracy of relative velocity measurements (2×10^{-4}) provide an overall precision in C much better than 1 % whatever the porous media and the fluid couples are. The used fluid couples are nonane-hexane, water-ethanol mixtures (40 % and 60 %) and glycerol-water mixtures; these couples allow us to vary continuously the viscosity ratio ($M = \mu_D/\mu_I$) from 1.4×10^3 to 7×10^{-4}. For immiscible displacement water is the wetting phase and hexane and nonane are the non-wetting phases. In the experimental procedure the sample is fully saturated with a vacuum impregnation technique. For each medium, each viscosity ratio and each flow rate, we record the time and space dependence C (x,t) of the concentration in our ten cross-sections of the sample.

RESULTS AND DISCUSSION

1. - One phase flow dispersion

In order to analyze our data, we can remark that, as we neglect the transverse dispersion in the one-dimension convection diffusion equation (1), this one is a function of the single variable $\xi = (x - Ut) / 2\sqrt{D_\parallel t}$. In figure 1, for four different flow rates U, we plot the concentration C versus ξ; data points correspond to different abscissa, x, from the inlet and different times t; the flow dependent dispersion coefficients $D_\parallel$ have been determined in order that our experimental data fit the theoretical solution of equation (1); this solution, $C(\xi) = \frac{1}{2} (1 - \text{erf } \xi)$, where erf is the error function, corresponds to the line curve in figure 1. This gaussian dispersion fits well our data for both consolidated and unconsolidated media whatever the flow rate and the viscosity ratio (M < 1) are. The corresponding dispersion coefficients $D_\parallel$ are given in figure 2 in a universal representation, $D_\parallel / U \cdot \ell_c$ versus a Peclet number $U \cdot \ell_c / D_m$, which allows a comparison between samples of different pore size ℓ_c. For glass beads, we get the two expected regimes: at low flow rates $D_\parallel$ becomes constant (slope -1 in figure 2) and equal to $D_m/1.7$; this

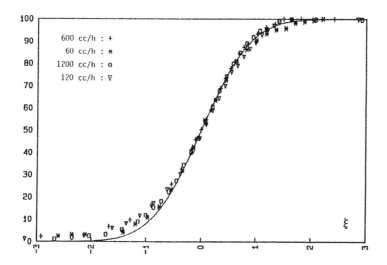

Figure 1. Concentration versus the reduced variable $\xi = \dfrac{x - Ut}{2\sqrt{D_\parallel t}}$ for different flow rates; x is the position from the inlet, t the time and U the flow rate. The dispersion coefficient D has been adjusted in order to fit the error function ($\frac{1}{2}(1 - \text{erf }\xi)$ line curve) solution of the convection-diffusion equation (1).

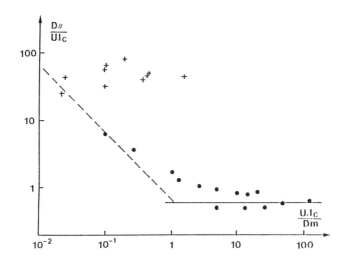

Figure 2. One phase flow dispersion: universal log-log plot of the reduced dispersion coefficient $D_\parallel/U\cdot\ell_c$ versus a Peclet number $U\cdot\ell_c/D_m$. $D_\parallel$ is the dispersion coefficient, U the flow rate, ℓ_c the characteristic pore size of the porous media, D_m the molecular diffusion coefficient. (● 200 μm glass beads, + sandstone). The dashed line has a slope -1 ($D_\parallel \sim D_m$) and corresponds to the zero flow rate regime. The full line corresponds to the mechanical dispersion regime ($D_\parallel \propto U$).

value of α is in reasonable accordance with our conductivity and acoustic measurements[12]. At high flow rates $D_{//}/U \cdot \ell_c$ becomes nearly constant and equal to roughly 0.5 . Broken glass sample (ϕ = 45 %, $\ell_c \sim$ 100 µm) exhibits a comparable behavior, but in fact these unconsolidated media have nearly the same porosity and the same topology ($\ell_c \sim$ d). The typical dispersion length is then $\ell_D \sim \ell_c$ (cf Table 1).

In sandstones, mechanical dispersion is also gaussian but $D_{//}$ increases slightly faster than U (figure 2) ; this feature can be explained by, either stagnation points (U $\cdot$ ln U), or boundary layer ($U^{4/3}$) or the power law of Sahimi et al ($D_{//} \sim U^{1.28}$), or the shoulder of de Archangelis et al[26] ; we cannot seriously distinguish between all these non-drastic laws because of experimental accuracy. But we do emphasize that, in contrast with unconsolidated media, the typical dispersion length ℓ_D (table 1) is roughly two-order of magnitude larger than ℓ_c ($\ell_D \sim$ 40 ℓ_c). Moreover, our universal plot, clearly demonstrates that in such a disordered porous medium the dispersion is enhanced compared to that of monodisperse beads. Our figure 2 surprisingly looks like the 2D-numerical simulations of disorder effect on dispersion (figure 2 of reference 26). The tortuosity value (α = 7) measured from conductivity is not large enough to account for the huge dispersion length we get ; would that mean that the involved tortuosity is a larger one such as the α_a of an ant in a porous maze ? Unfortunately we do not have, at the moment, a measured value of this tortuosity.

2. - Two phase flow dispersion

The use of porous medium saturated with two immiscible fluids allows us to change the structure of the subnetwork on which diffusion occurs, without changing the typical characteristic size ℓ_c : dispersion would then be affected through tortuosity changes. For drainage process, we first fully saturate the sandsstone with wetting fluid (non-wetting fluid saturation S_0 = 0). We inject non-wetting fluid at low flow rate up to reach a steady state with a certain repartition of fluids along the sample ; then we inject another miscible non-wetting fluid which disperses on the non-wetting subnetwork we have created through drainage, without changing the complementary wetting network. Then thanks to a new injection of non-wetting fluid at a higher flow rate, we get another repartition of fluids further away from the drainage percolation threshold (S_0 = 0) on which we study the dispersion and so on...

In our sandstones, for fluid saturations obtained through successive drainage processes, three main features of the dispersion on the non-wetting phase subnetwork are observed :

i) At the same flow rate U, the longitudinal dispersion is enhanced compared to one phase flow and leads to huge dispersion length ($\ell_D \sim$ 200 $\ell_c \sim$ 3 mm).

ii) The dispersion coefficient is affected by changes of the non wetting phase saturation : the closer to the drainage threshold (S_0 = 0), the larger the dispersion coefficient. Figure 3 quotes this divergence in a plot of $D_{//}/U \cdot \ell_c$ versus the non-wetting phase saturation S_0.

iii) The dispersion is no more gaussian : concentration profiles exhibit a long tail. This long tail can be well fitted by an exponential decay of characteristic time τ . The dispersion coefficient and τ simultaneously diverge as the drainage percolation threshold is reached (figure 4).

All these features clearly show the influence of the morphology of the network in which dispersion occurs. Theoretical 3D-predictions on the divergence of the dispersion coefficient in the vicinity of percolation threshold crucially depend on the assumed structure of backbone and dead-ends. The three lines in figure 3 correspond

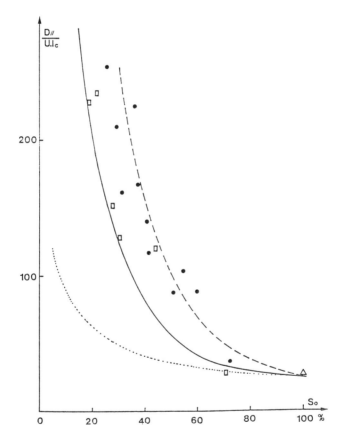

Figure 3. *Two phase flow dispersion in sandstone : variation of the reduced dispersion coefficient ($D_{\parallel}/U \cdot \ell_c$) on the non-wetting phase versus the non-wetting phase saturation S_0 for two different injection processes : ● 1 cc/h , □ 2 cc/h . The lines correspond to theoretical predictions on the divergence of the dispersion coefficient in the vicinity of the percolation threshold ($S_0 = 0$) : de Gennes (...), Vannimenus (---) and Sahimi et al (———). △ corresponds to the one-phase flow experiment.*

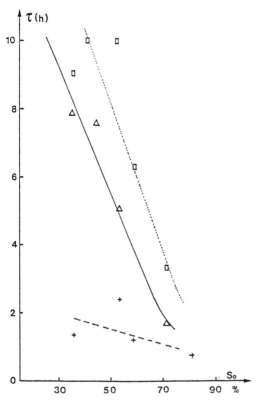

Figure 4. Two-phase flow dispersion in sandstone : variations of the characteristic time τ of the exponential decay of the long tail of anomalous dispersion versus non-wetting saturation S_0. The lines through data are only guides to follow τ-variations for three different injection processes : from top to bottom (..., ——, ---) the injection flow rate of the drainage process (1 cc/h, 2 cc/h, 4 cc/h) increases leading to larger S_0 values.

respectively to the weak de Gennes divergence[5] (exponent - 0.5), the larger Vannimenus one[27] (exponent - 2) and the Sahimi et al[6] calculations. Our data are in reasonable agreement with either Sahimi et al or Vannimenus predictions. This agreement would mean that the observed divergence is mainly due to the divergence of the backbone tortuosity : the backbone consists of a complicated structure of mazes which provide alternative flow passages and links connecting the mazes. The long tail corresponds to anomalous dispersion and was theoretically expected[6] ; this tail only slightly affects dispersion on the backbone and corresponds to molecular diffusion into the dead-ends ; from our τ values (figure 4) one can deduce that the typical diffusion length ($\sqrt{2 D_m \tau} \sim 4$ mm for $\tau \sim 5$ h) is very close to the dispersion length ℓ_D : this would mean that the structure of the dead-ends is similar to the structure of the backbone mazes. Last feature of this experiment, as we increase the non-wetting fluid concentration through a higher flow rate drainage, dispersion coefficients decrease slightly (full circles and squares in figure 3) ; in the same time the long tail corresponds to less and less allowed porosity (from 25 % downwards) and the characteristic time τ decreases drastically (figure 4). These observations are consistent with an analogous structure of maze in backbone and dead-ends : as the oil content increases, dead-ends become more and more connected to the backbone without affecting $D_{//}$ and the dead-ends become smaller and smaller (τ decreases).

In order to summarize our results on both one or two phase flow mechanical dispersion we can propose a simple approximation for the dispersion coefficient :

$$D_{/\!/} = \alpha U \ell_c \qquad (5)$$

where ℓ_c is the typical pore size deduced from permeability and conductivity measurements, U is the mean flow velocity measured during the miscible displacement and α is the dispersion tortuosity which is the backbone tortuosity for two phase flow, and has to be compared to the ant tortuosity for one phase flow.

3. - Unstable miscible displacements

When the injected fluid is less viscous than the displaced one, the observed profiles are unstable, no more gaussian-like whatever the porous medium and the flow rates are, even for viscosity ratios, close to 1. Figure 5 gives a series of profiles (concentration versus distance x from the inlet) for different viscosity ratios $M = \mu_D/\mu_I$ at the same low flow rate and for the same quantity of injected fluid : as M increases the gaussian-like piston profile becomes more and more extended onwards. 2D-numerical simulations of Sherwood and Nittmann[16] qualitatively imitate our observed profile. This might encourage 3D numerical simulations in the future. In our data, the larger M, the smaller the fluid concentration needed to induce the breakthrough ; the sweep efficiency, corresponding to this precursor concentration, is given versus M in figure 6 : as M increases the sweep efficiency tends towards a constant ($\lesssim$ 10 %) which would be the DLA limit (M = ∞) in 3D, for our sample size. These results are

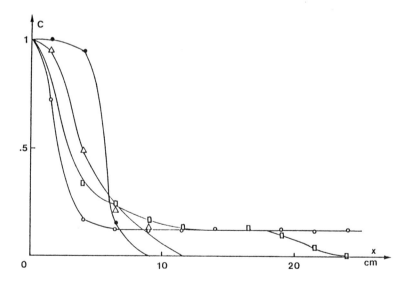

Figure 5. *Unstable miscible displacement profiles : concentration of less viscous fluid versus distance x from the inlet for the same quantity of injected fluid. Lines through the data are only a guide ; the corresponding viscosity ratios* $M = \mu_D/\mu_I$ *, of the displaced fluid* (μ_D) *to the injected one, are M = 2.5* ($\bullet$) *, 11* ($\triangle$) *, 110* ($\square$) *and 1400* ($\circ$) *; the porous medium is a sandstone.*

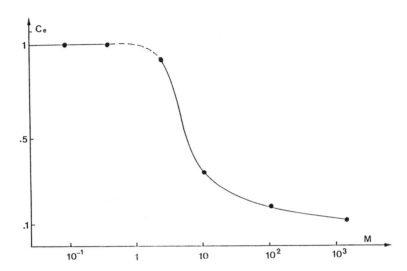

Figure 6. *Variation of the sweep efficiency C_e, concentration of injected fluid at the breakthrough, versus the viscosity ratio $M = \mu_D/\mu_I$ of the displaced fluid to the injected fluid in sandstone.*

analogous to the 2D five spot simulations of De Gregoria[29]. Last feature of this experiment is the influence of the flow rate ; figure 7 gives, for broken glass, profile variations at different flow rates U for the same quantity of injected fluid and a viscosity ratio $M = 11$:

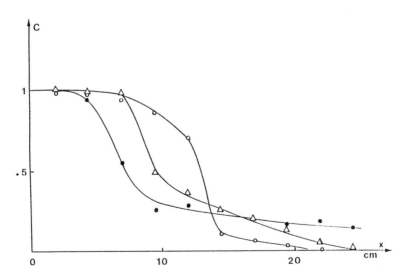

Figure 7. *Unstable miscible displacement profiles for a same quantity of injected fluid and for the same viscosity ratio $M = 11$ in broken glass. Lines through the data are only a guide for the different flow rates : 2 cm³/h (○) , 8 cm³/h (△) and 32 cm³/h (●).*

as U increases, the number of growing fingers increases. We have to notice that the flow velocity dependence decreases as M increases : this means that in the DLA limit (M = ∞) of an injected immiscible fluid, the flow rate has no influence as expected.

CONCLUSION

Thanks to our acoustical technique, we have measured concentration profiles during miscible fluid displacements through both consolidated and unconsolidated porous media at different flow rates and viscosity ratios.

When the displacement is stable we have analyzed the hydrodynamic dispersion. In disordered medium our measurements lead to a large dispersion length which is mainly due to the tortuosity of the sandstone.

With the help of sandstone saturated with another immiscible fluid, we have changed the structure of the network on which dispersion occurs ; our data are in reasonable agreement with theoretical predictions on the divergence of the dispersion coefficient in the vicinity of a percolation threshold ; the divergence is due to the tortuous nature of the backbone.

When the displacement is unstable we have determined the main features of the viscous fingering with viscosity ratios and flow rate variations. 2D numerical simulations imitate our experimental data. This leads to our final question to theoreticians in order to extract, during calculations, much more data comparable to experiments, such as density profiles, sweep efficiency and as far as possible in the real 3D space.

REFERENCES

1. G. I. Taylor, *Proc. Roy. Soc.* **A 219**, 186 (1953).
2. P. G. Saffman, *J. Fluid Mech.* **6**, 321 (1959) ; *ibid.* **7**, 194 (1960).
3. P. G. Saffman and G. I. Taylor, *Proc. Roy. Soc. London* **A 245**, 312 (1958).
4. R. L. Chuoke, P. van Meurs and C. van der Poel, *Trans. Am. Inst. Min. Mett. Pet. Eng.* 216, 188 (1959).
5. P. G. de Gennes, *J. Fluid Mech.* **136**, 189 (1983).
6. M. Sahimi, H. T. Davis and L. E. Scriven, *Chem. Eng. Comm.* **23**, 329 (1983) ; M. Sahimi, B. D. Hughes, A. L. Heiba, H. T. Davis and L. E. Scriven, *Chem. Engr. Sci.* (to be published).
7. L. Paterson, *Phys. Rev. Lett.* 52, 1621 (1984) ; *Phys. Fluids* 28, 26 (1985).
8. T. A. Witten and L. M. Sander, *Phys. Rev.* **B 27**, 5686 (1983).
9. H. O. Pfannkuch, *Rev. de l'Institut Français du Pétrole* 18, 215 (1963).
10. J. J. Fried and M. A. Combarnous, *Advances in Hydroscience* 7, 169 (1971) ; J. Bear, *Dynamics of Fluids in Porous Media* (Elsevier, 1971) ; A. E. Scheidegger, *The Physics of Flow in*

Porous Media (University of Toronto Press, 1974).
11. D. L. Koch and J. F. Brady, *J. Fluid Mech.* **154**, 399 (1985).
12. J.-C. Bacri, C. Leygnac and D. Salin, *J. Physique Lett.* **45**, L-767 (1984).
13. E. Charlaix, J.-P. Hulin and T. J. Plona, *C. R. Acad. Sc. Paris* (to be published).
14. R. Lenormand and C. Zarcone, in *Proc. of the Vth Int. Meeting Phys. Chem. Hydro.*, Tel-Aviv (Dec. 1984).
15. J. Nittmann, G. Daccord and H. E. Stanley, *Nature* **314**, 141 (1985).
16. J. D. Sherwood and J. Nittmann, *J. Physique* **47**, 15 (1986).
17. J.-C. Bacri, C. Leygnac and D. Salin, *J. Physique Lett.* **46**, L-467 (1985).
18. R. Aris, *Proc. Roy. Soc. London* **A 219**, 186 (1953).
19. G. de J. de Jong, *Trans. Amer. Geophys. Union* **39**, 67 (1958).
20. P. G. de Gennes, *La Recherche* **7**, 919 (1976).
21. C. Mitescu, H. Ottavi and J. Roussenq, *AIP Conf. Proc.* **40**, 377 (1979).
22. Y. Gefen, A. Aharony and S. Alexander, *Phys. Rev. Lett.* **50**, 77 (1983).
23. J. G. Berryman, *Appl. Phys. Lett.* **37**, 382 (1980).
24. D. L. Johnson, T. J. Plona, C. Scala, F. Pasierb and H. Kojima, *Phys. Rev. Lett.* **49**, 1840 (1982).
25. R. G. Larson and H. T. Davis, *J. Phys. C, Sol. St. Phys.* **15**, 2327 (1982).
26. L. de Arcangelis, J. Koplik, S. Redner and D. Wilkinson, *Phys. Rev. Lett.* **57**, 996 (1986)
27. J. Vannimenus, *J. Physique Lett.* **45**, L-1071 (1984).
28. R. Lenormand, *C. R. Acad. Sc. Paris* **301**, 247 (1985).
29. A. J. De Gregoria, *Phys. Fluids* **28**, 2933 (1985).
30. A. J. Katz and A. H. Thompson, submitted to Phys. Rev. Lett. ; submitted to J. Geophys. Research.
31. D. M. Johnson, J. Koplik, L. Schwartz, submitted to Phys. Rev. Lett.

THE TRANSPORT PROPERTIES OF NON-DILUTE SUSPENSIONS. RENORMALIZATION VIA AN EFFECTIVE CONTINUUM METHOD

A. Acrivos and E. Y. Chang
Department of Chemical Engineering
Stanford University, Stanford, CA 94305-5025

ABSTRACT

In this paper, we present a new method for resolving the non-absolutely convergent and divergent integrals which often arise in the course of calculating the effective transport parameters of two-phase materials. Although several well-established techniques for accomplishing this have already been discussed previously in the literature, this new method appears to be more direct and to have a stronger physical basis than those previously proposed. Briefly, the main feature of the method is that it replaces part of the composite in the system with an effective continuum having the bulk property being calculated. By means of this step, long range interactions between particles are accounted for in a more complete way, and as a result, the convergence difficulties are resolved.

To illustrate its use, this method is applied to three different problems which involve the calculation of:
1) The effective thermal conductivity of a randomly distributed isotropic, and stationary suspension of identical spheres to $O(c^2)$.
2) The sedimentation rate of a dilute random suspension of identical solid spheres settling under the influence of gravity to $O(c)$.
3) The effective viscosity of a random well-mixed suspension of solid spheres in a Newtonian fluid to $O(c^2)$.

Finally, it is pointed out that this effective continuum method is useful for estimating such transport quantities on its own without resorting to detailed many body calculations, and that, as shown in another publication, it can be extended to higher orders in the volume fraction of the dispersed phase by means of a self consistent scheme.

1. INTRODUCTION

This paper deals with the problem of calculating the effective properties of a two-phase composite, in which one of the constituents is randomly dispersed within the other, given the properties of the pure materials as well as the geometry, spatial arrangement and concentration of the inclusions. This problem is of course an old one which has been studied off and on for over a century.

Consider as an example the case of a two-phase material composed of equal-sized spheres, of radius a, henceforth set equal to unity, thermal conductivity α, and volume concentration c, which are imbedded in a pure matrix having unit thermal conductivity. The material is placed in a temperature field which, on a length scale much larger that the microscale a, has a uniform gradient G_i. For an isotropic two-phase composite, the bulk heat flux Q_i is then a scalar multiple of G_i, i.e.

$$Q_i = -k^* G_i \qquad (1)$$

where k^* is the effective thermal conductivity of the system. It can be shown rigorously[1] that,

$$Q_i = -G_i - (\alpha - 1) c \langle (\frac{\partial T}{\partial x_i})_0 \rangle \qquad (2)$$

where $(\partial T/\partial x_i)_0$ is the temperature gradient at the center of a given sphere, the so-called test sphere, and the brackets denote the conditional average of this quantity over the positions of all the other spheres in the sample given that the test sphere is kept fixed with its center at the origin. Thus, since $\langle (\partial T/\partial x_i)_0 \rangle$ is a scalar multiple of G_i for a material which is isotropic and spatially homogeneous on the macroscale, eqs. (1) and (2) provide a formal vehicle for evaluating k^* as long as $\langle (\partial T/\partial x_i)_0 \rangle$ can first be determined. This can easily be accomplished for an infinitely dilute suspension where it suffices to compute $(\partial T/\partial x_i)_0$ at the center of a single sphere when the temperature gradient at infinity equals G_i. As is well-known

$$\langle (\frac{\partial T}{\partial x_i})_0 \rangle \rightarrow -\frac{3}{\alpha + 2} G_i \quad \text{as} \quad c \rightarrow 0 ,$$

and hence

$$k^* = 1 + 3\beta c + O(c^2) , \quad \beta \equiv \frac{\alpha - 1}{\alpha + 2} \qquad (3)$$

a result first due to Maxwell.

The extension of the analysis to $O(c^2)$ is, however, not so straightforward because, if one supposes that the $O(c^2)$ term in eq. (3) arises from the presence of a second sphere at $\mathbf{x} = \mathbf{R}$ whose extra contribution to $(\partial T/\partial x_i)_0$ is first multiplied by a probability distribution function and then integrated over the interval $2 \leq R \leq \infty$, the resulting integral is only conditionally convergent since the integrand decays as R^{-3}. The reason for the failure of this approach lies in the fact that, although the probability of finding a third particle in the vicinity of the other two is indeed only $O(c^2)$, and therefore negligible to this order of approximation if R is $O(1)$, the influence on $(\partial T/\partial x_i)_0$ of all the other particles in the suspension cannot be neglected as $R \to \infty$ because, although their individual contribution is only $O(R^{-3})$, their combined effect is $O(c)$ owing to their very large number.

As is well-known, several methods have already been proposed for eliminating this difficulty of which we shall briefly discuss only three. The first, due to Batchelor[2], makes use of the fact that there often exists a quantity, which has the same far field dependence on R as the term under consideration, in this case $(\partial T/\partial x_i)_0$, and whose bulk value is known from the statement of the problem. By subtracting a suitable multiple of this quantity from $(\partial T/\partial x_i)_0$ and then averaging, a convergent integral is thus obtained which requires for its determination the detailed two-particle solution for the problem at hand. Unfortunately, although often successful, this renormalization technique occasionally fails and at least in one case leads to a non-unique result[3].

The method due to O'Brien[4] begins by treating the composite as being enclosed by finite "macroscopic" boundaries on which the relevant parameters assume their bulk values. A formal solution is then constructed for the appropriate system of equations and the quantity of interest is computed. Although the technique is rigorous, its application requires that the Green's function for the particular problem be known.

The most general and systematic approach to such calculations is undoubtedly that due to Hinch[5] which aims to derive the equations that govern locally the various average field quantities by taking suitable ensemble averages of the conservation laws and constitutive equations. Unfortunately, the drawback of this method is its mathematical complexity.

In what follows, we shall present yet another renormalization technique which removes the divergences referred to above rather more quickly and directly than the methods developed thus far and which appears to provide a somewhat clearer physical insight into some of the main features of this whole subject than can be gleaned from the earlier analyses. Thus, although no new results will be presented on this occasion, it is hoped that the technique to be described will be found to be of some interest especially since it can form the basis for estimating the effective properties even in concentrated systems[6].

2. THE EFFECTIVE THERMAL CONDUCTIVITY TO $O(c^2)$

We return to the problem, mentioned in the introduction, of calculating the effective thermal conductivity of a stationary, random and statistically homogeneous suspension of equal-sized spherical particles in a matrix of uniform conductivity equal to unity. In view of eqs. (1) and (2), this requires the determination of $(\partial T/\partial x_i)_0$, the temperature gradient at the center of the test

particle, and its average, given the bulk temperature gradient G_i.
Now, in a dilute suspension, the test particle is surrounded mostly
by pure matrix up to a distance R, which is comparable to the
particle radius since the probability of finding a second sphere in
such a neighborhood is $O(c)$. On the other hand, a spherical surface
of radius R will cut through a large number of particles even if
$c \ll 1$, provided that R is made sufficiently large. In other words,
it is clear that, on a length scale $R \gg 1$, the suspension acts like
an effective medium with conductivity k^*, so that the test sphere
can be viewed as being surrounded by pure matrix in its immediate
neighborhood, and by a uniform medium with a conductivity equal to
k^* at $R \gg 1$. It follows that an intermediate region will exist
within which the conductivity will vary continuously from unity to
k^*. The precise details of this intermediate region, although
certainly significant, are not crucial in resolving the
aforementioned non-absolute convergence difficulties.

An approximate way of proceeding is by means of the self-consistent scheme described by Hashin[7] in which the test sphere is viewed as being embedded in an effective continuum of conductivity k^* but separated from it by a spherical shell of pure matrix of radius Γ. The solution of the appropriate heat conduction problem then yields:

$$T = A\, G_j\, x_j \quad , \quad 0 \leq r \leq 1 \tag{4a}$$

$$T = \left(B + \frac{C}{r^3} \right) G_j\, x_j \quad 1 \leq r \leq \Gamma \tag{4b}$$

$$T = \left(1 + \frac{D}{r^3} \right) G_j\, x_j \quad r \geq \Gamma \tag{4c}$$

where the coefficients A, B, C and D satisfy

$$A = B + C \; ; \quad \alpha A = B - 2C \tag{5a}$$

$$B + \frac{C}{\Gamma^3} = 1 + \frac{D}{\Gamma^3} \; ; \quad B - \frac{2C}{\Gamma^3} = k^* \left(1 - \frac{2D}{\Gamma^3} \right) \tag{5b}$$

Moreover, in view of eqs. (1) and (2),

$$k^* = 1 + (\alpha - 1)\, c\, A \tag{6}$$

which in conjunction with (5) leads to the desired expression for k^* as a function of c. Expanding to $O(c^2)$, we obtain

$$k^* = 1 + 3\beta c + 3\beta^2 c^2 \left(1 + \frac{2\beta}{\Gamma^3} \right) + O(c^3) \tag{7}$$

Several features of this solution are worth noting at this point. First of all, if eqs. (5) and (6) are solved and then Γ is made to increase without bound, the resulting expression for k^* becomes

$$k^* = 1 + 3\beta c + 3\beta^2 c^2 + O(c^3) \qquad (8)$$

which, to $O(c^2)$, is the Hashin-Shtrikman lower (upper) bound for $\beta > 0$ ($\beta < 0$). On the other hand, letting $\Gamma \to \infty$ in (5) <u>ab initio</u> leads to a degenerate system. Thus, the existence of an effective continuum for $r \geq \Gamma$ constitutes an essential component of this model even though, to $O(c^2)$, Γ can be made arbitrarily large. Note that[7] setting $\Gamma^{-3} = c$ gives the Hashin-Shtrikman lower (upper) bound to all orders in c, i.e.

$$k^* \genfrac{}{}{0pt}{}{>}{<} \frac{1 + 2\beta c}{1 - \beta c} \qquad (9)$$

At the other extreme, $\Gamma = 1$, one obtains the implicit expression first derived by Landauer[8],

$$k^* = 1 + \frac{3(\alpha - 1)ck^*}{\alpha + 2k^*} = 1 + 3\beta c + 3\beta^2 c^2 (1 + 2\beta) + O(c^3) \qquad (10)$$

which clearly gives an upper (lower) bound for k^*.

It is important to appreciate at this stage that the use of an effective continuum for $r \geq \Gamma$ is equivalent to replacing the effect of all the spheres, whose centers lie beyond $r = \Gamma$, with that of only the dipoles at their centers, whose strengths are induced only by the incident bulk temperature field plus the first reflection disturbance arising from the presence of the test sphere. Since the strength of the dipoles associated with this contribution is proportional to the volume of each sphere, the associated dipole moment per unit volume is independent of their size. Consequently, Landauer's result, eq. (10), corresponds to the case in which the test sphere is surrounded by a composite containing spheres of infinitesimally small size so that their centers could range over the whole domain $1 \leq R \leq \infty$. When all the spheres are of the same size as the test sphere, however, and all are impenetrable, it is clear that R must lie beyond 2, and hence, for a well mixed suspension, one should subtract from eq. (10) the contribution arising from a uniform distribution of dipoles within the space $1 \leq R \leq 2$. For a dilute suspension, this calculation can be performed as follows:

Let the test sphere have its center at the origin and let us place at $\mathbf{x} = \mathbf{R}$ a small sphere of radius $\epsilon \ll 1$. The external temperature field T^E incident on this small sphere consists of two terms: the undisturbed temperature $G_j x_j$ and the first reflection from the test sphere, hence

$$T^E = G_j x_j \left(1 - \frac{\beta}{|x|^3}\right) \qquad (11)$$

Consequently, the temperature disturbance induced by the small sphere becomes

$$-\varepsilon^3 \beta \frac{y_j}{|y|^3} \left(\frac{\partial T^E}{\partial x_j}\right)_{x=R} , \qquad y \equiv x - R \qquad (12)$$

which in turn contributes to $(\partial T/\partial x_i)_0$ an amount equal to

$$-\frac{3\varepsilon^3 \beta}{\alpha+2} G_j \left\{ \frac{1}{R^3} \left(\delta_{kj} - \frac{3R_k R_j}{R^2}\right) \left[\delta_{ik} - \frac{\beta}{R^3}\left(\delta_{ik} - \frac{3R_i R_k}{R^2}\right)\right]\right\} \qquad (13)$$

where use has been made of eq. (11). For a well mixed suspension, the number density of small spheres equals $3c/4\pi\varepsilon^3$ hence, integrating eq. (13) over the spherical shell $1 \le R \le 2$ gives, in conjunction with eq. (1) and (2), a contribution to k^* equal to

$$18 \beta^3 c^2 \int_1^2 \frac{dR}{R^4} = \frac{21}{4} \beta^3 c^2 . \qquad (14)$$

We point out that the integral of eq. (13) over the spherical shell is no longer beset by convergence difficulties since it is taken over a finite region. Subtracting eq. (14) from the right-hand side of eq. (10) therefore leads to

$$k^* = 1 + 3\beta c + 3\beta^2 c^2 \left(1 + \frac{\beta}{4}\right) + O(c^3) \qquad (15)$$

which, as expected, is identical with eq. (7) if ε is set equal to 2.

The coefficient of the $O(c^2)$ term in eq. (15) is of course only approximate because it does not take fully into account the interaction between the test sphere and that whose center is located at R. A more complete calculation requires therefore the detailed solution of a two sphere conduction problem. Indeed, from Jeffrey's[1] eq. (6.4) we have that

$$k^* = 1 + 3\beta c + 3\beta^2 c^2 \left\{1 + \frac{\beta}{4} + \frac{3\beta}{16}\frac{\alpha-1}{2\alpha+3} + \frac{\beta^2}{64} + \ldots\right\} + O(c^3) \qquad (16)$$

where the terms in the brackets beyond the second arise from the higher order reflections between the two spheres.

We see then that according to the procedure just described, the coefficient of the $O(c^2)$ term in the expression for k^* can simply be obtained as follows:

1. The test particle is viewed as being surrounded by a spherical shell of pure matrix within the space $1 \leq R \leq 2$, and by a continuum with effective thermal conductivity k^* for $R > 2$. A single-sphere heat conduction problem is then solved and the corresponding contribution to k^* is evaluated. Such a procedure automatically eliminates any divergent or conditionally convergent integrals and is entirely equivalent to the renormalization methods proposed earlier.

2. To improve the accuracy of the expression for k^* for $O(c^2)$, it is necessary to solve a two-sphere conduction problem from which, however, the contributions to k^* arising from the first two reflections must first be subtracted since these have already been accounted for under #1.

Although a careful comparison between the procedure just outlined and the methods proposed earlier in the literature[4,5] will indicate that all of them are entirely equivalent, it is felt that the use of an effective medium approximation eliminates the need for renormalizing non-absolutely convergent integrals in a more direct and physical manner than is achieved otherwise. Indeed by taking advantage of the fact that the suspension far from the test sphere behaves as an effective continuum whose influence cannot be viewed as being simply the average of two body interactions, this method automatically accounts for complex interactions such as "shielding". In addition, as has been shown elsewhere[6], this effective medium description can be extended to more concentrated systems and yields an approximate but accurate expression for k^* up to $c = 0.60$.

We also note parenthetically that if, in the effective medium calculation, the test sphere is surrounded by a continuum having a position dependent effective thermal conductivity $\kappa = 1 + (k^*-1)\Psi(r,c)/c$, where $\Psi(r,c)$ is the probability that a given point outside the test sphere will lie within one of the other spheres comprising the suspension, then the coefficient multiplying the term $3\beta^2 c^2$ in eq. (16) becomes

$$1 + 0.42135 \beta$$

which is exact to $O(\beta)$ as can be shown either analytically[6] or by comparison with Jeffrey's numerical results[1].

3. THE SEDIMENTATION RATE OF A DILUTE SUSPENSION OF SPHERES TO $O(c)$

Another class of systems which may be analyzed using these methods is that involving the motion of suspensions. Recall that, even when the Reynolds number is small, and therefore inertial forces are negligible, the interactions between particles suspended in a moving fluid are still complex, and consequently most problems of this type are treated by isolating such interactions to groups of only a few particles. As we have seen in the course of calculating the effective thermal conductivity of a dispersion of spheres, this can lead to convergence difficulties. Again, however, we may circumvent these by applying our effective continuum method and replacing part of the suspension with an effective fluid having a specified viscosity and density.

We begin by calculating the sedimentation rate to O(c) of a suspension of solid spheres in a viscous fluid. If all the spheres in the suspension are identical, this is equivalent to finding the fall speed of any one of them. In addition, when the suspension is extremely dilute, and there are no interactions between any of the settling particles, each sphere falls with its corresponding Stokes settling velocity,

$$U_0 = \frac{2}{9} \frac{(\rho_s - \rho_f)g}{\mu a^2} \ . \tag{17}$$

Thus, if the volume fraction of spheres in the suspension is larger, but still small, it seems reasonable to correct the settling velocity of one sphere, which we shall call the test sphere, by simply accounting for the presence of another one and then integrating over all its possible locations. Unfortunately, the effect of the second sphere on the settling speed of the first is approximately that of a Stokeslet when their separation distance R is large, and is therefore O(1/R). The integral of this correction to the fall speed over the configurational ensemble is therefore not merely non-absolutely convergent, as was the case in the previous problem, but it is actually divergent. Again though, by using the techniques referred to in the introduction, it is of course possible to modify this approach and obtain a rigorous solution involving only convergent integrals[2,5].

Let us, however, remove the divergence caused by the influence of the particles at infinity on the test sphere, by applying the effective continuum method in nearly the same fashion as was done in the previous problem. Without loss of generality, we set the radii of the spheres, the density and viscosity of the clear fluid, and the Stokes settling rate, U_0, of the spheres to unity. As a result, the density of the spheres is $1 + 9/2g$. Furthermore, on account of continuity, the suspension is, on the average, stationary at infinity. Thus, according to our effective continuum method, we surround the test sphere with an effective continuum with density $1 + 9c/2g$ and viscosity $1 + 5c/2$, the latter given by Einstein's well-known expression, except for the concentric layer from $1 \leq r \leq 2$, which contains clear fluid. The increased density reflects the presence of a force exerted on the fluid due to the Stokeslet at each sphere while the increased viscosity reflects the extra stress exerted by the corresponding stresslet. Unfortunately, there exists one additional singularity not accounted for in this continuum treatment. Specifically, the correction to the settling speed of the test sphere brought about by this singularity, referred to by Hinch[5] as the force quadrapole, is given by the second term of the expression for the disturbance velocity of a sphere of unit radius translating with uniform velocity **U**

$$u_i = \frac{3}{4} U_k \left(\frac{\delta_{ik}}{y} + \frac{y_i y_k}{y^3} \right) + \frac{3}{4} U_k \left(\frac{\delta_{ik}}{3y^3} - \frac{y_i y_k}{y^5} \right) \tag{18}$$

where y_i is the position vector referred to the center of the second sphere. Since this second term decays as R^{-3}, it is clear that the corresponding conditional average which assess the influence of all the spheres on the settling speed of the test sphere, is obviously non-absolutely convergent. As a result, this must be accounted for before the two sphere solution is averaged.

To begin with, we note that if we replace the disturbance velocity induced by each sphere in a dilute suspension by placing a Stokeslet at its center, henceforth referred to as the pointwise distribution, we recover only the first term of eq. (18). By distributing Stokeslets uniformly over the volume of the same sphere, however, we obtain

$$u_i = \frac{3U_k}{16\pi} \int_{|\mathbf{y}'|\leq 1} \left(\frac{\delta_{ik}}{s} + \frac{s_i s_k}{s^3} \right) dV(\mathbf{y}') \tag{19}$$

where $\mathbf{s} = \mathbf{y} - \mathbf{y}'$, $\mathbf{y}$ is the position vector, and $\mathbf{y}'$ is the integration vector whose origin is at the center of the second sphere, which in turn is at **R**. On evaluating eq. (19), we find that

$$u_i = \frac{3}{4} U_k \left[\left(\frac{\delta_{ik}}{y} + \frac{y_i y_k}{y^3} \right) + \frac{3}{5} \left(\frac{\delta_{ik}}{3y^3} - \frac{y_i y_k}{y^5} \right) \right] \tag{20}$$

Consequently, we can recover eq. (18) by means of a linear combination consisting of -2/3 of the pointwise Stokeslet distribution added to 5/3 of the volume distribution. Furthermore, to obtain the correction to the settling speed of the test sphere due to this modified effective continuum, it suffices, because of linearity, to consider the correction from each distribution separately, and then combine them according to the above ratios to give the complete result.

We point out that we may compute separately the effects of viscosity and density since their mutual interaction is $O(c^2)$ and therefore negligible for this order of approximation. Proceeding along these lines we first apply the effective continuum method to estimate the effects of density alone on the settling speed of the heavy sphere, which in turn involves calculating the correction to the fall speed of a solid sphere of unit radius, settling in an effective fluid which has unit viscosity and, for the pointwise distribution, a density equal to unity in the spherical region $1 \leq r \leq 2$ and equal to $1 + 9c/2g$ thereafter. The problem is axisymmetric, and therefore is described[8] by a partial differential equation for the streamfunction outside the sphere, where account must be taken for the body force due to the density difference between the fluids across the surface r = 2. This equation, in turn, is separable and may be further reduced to an ordinary differential equation which is then solved subject to the boundary conditions of no-slip at the surface of the test sphere, continuity of velocity and stress at r = 2, and vanishing of the velocity at infinity, to yield,

$$U = 1 - \frac{11c}{2} \qquad (21)$$

where U refers to the settling speed divided by U_0 as given in eq. (17). The corresponding expression for the volume distribution, which is obtained in a manner analogous to that outlined above but with a variable density equal to $1 + 9\Psi/2$ for $1 \leq r \leq 3$, where, as mentioned at the end of the previous section, Ψ is the probability that a given point lies within an inclusion[6], and requiring that the streamfunction and its first 3 derivatives in r be continuous at r = 3, is

$$U = 1 - \frac{26c}{5} \qquad (22)$$

The appropriate linear combination of eqs. (21) and (22) gives

$$U = 1 - 5c \qquad (23)$$

which is identical to the renormalization result given by Batchelor[2].

These corrections may also be derived individually by means of the singularity technique employed in the previous section. In the case of the volume distribution one first calculates the settling speed of a sphere which is completely surrounded by a fluid of density $1 + 9c/2g$, and then subtracts out the effect of the singularities in the exclusion region, here $1 \leq r \leq 3$, in proportion to the amount by which the inclusion volume fraction exceeds Ψ.

We now consider the effect of the change in viscosity on the settling speed of the test sphere. The corresponding exact system consists of a heavy sphere falling in a suspension of neutrally buoyant spheres of equal size. Applying the effective continuum model, we set the viscosity equal to $1 + 5c/2$ for $r > 2$ and unity for $1 \leq r \leq 2$, while the density is set equal to unity everywhere outside the test sphere. The boundary condition of no-slip is retained on the sphere's surface, the velocity vanishes at infinity, and continuity of velocity and stress is imposed at r = 2. The mathematical problem is again axisymmetric, and its solution yields, on account of the extra drag on the test sphere due to the increased viscosity beyond r = 2,

$$U = 1 - \frac{205}{128}c \ . \qquad (24)$$

Finally, combining eqs. (23) and (24) leads to the approximate correction to the sedimentation of a dilute suspension of identical spheres

$$U = 1 - 6.60\ c \qquad (25)$$

a result identical to that found by Hinch[5].

To obtain the exact $O(c)$ correction, one must of course include the effect of all the other reflections from the full two sphere solution, a task which is now straightforward since these remaining terms are too small to cause difficulties. As shown by Batchelor[2],

$$U = 1 - 6.55\ c \qquad (26)$$

for a well-mixed suspension.

4. THE EFFECTIVE VISCOSITY OF A SUSPENSION OF SPHERES TO $O(c^2)$

We now consider the last problem, namely that of finding the effective viscosity of a well-mixed suspension of spheres to $O(c^2)$. We mention in passing that although such a distribution does not account for the effect of two-particle trajectories, the assumption of well-mixedness adequately models the instantaneous behavior of an initially random suspension as well as the analogous problem of finding the effective elastic modulus of a composite containing spherical inclusions. In other words, in a suspension undergoing a straining motion for some time, the particles are not well mixed because the flow induces migration of particles relative to each other, thereby affecting the $O(c^2)$ coefficient of the effective viscosity. For more details on this, the reader is referred to Batchelor and Green's treatment of this problem[9] along with their calculation of the effective viscosity. As it turns out, however, the detailed form of the particle distribution function does not affect the difficulties associated with the presence of non-absolutely convergent integrals because any non-uniformities due to the interaction of two spheres are necessarily a local effect, as opposed to convergence problems in the conditional average integrals which arise from interactions at infinity.

Recall now that the effective viscosity of an isotropic suspension of particles may be defined as

$$\langle \sigma_{ij} \rangle - \frac{1}{3}\delta_{ij}\langle \sigma_{kk} \rangle = 2\mu^* \langle e_{ij} \rangle \qquad (27)$$

where σ_{ij} is the local stress tensor, e_{ij} is the rate of strain in the fluid, and the brackets $\langle\ \rangle$ denote once again the bulk average over all configurations of the inclusions. Moreover, for a statistically stationary, homogeneous and isotropic suspension, the effective viscosity satisfies[9]

$$n \langle S_{ij} \rangle = 2(\mu^* - 1) \langle e_{ij} \rangle \qquad (28)$$

where n is the number density of the spheres in the suspension and S_{ij} is the stresslet associated with one of the spheres, hereafter placed at the origin and referred to as the test sphere, and is defined by the integral

$$S_{ij} = \int_A (\sigma_{ik} x_j - \frac{1}{3} \delta_{ij} \sigma_{lk} x_l) n_k \, dA \qquad (29)$$

over its surface. According to Rallison[10], eq. (29) may be replaced by

$$S_{ij} = \frac{20\pi}{3} (E_{ij} + \frac{1}{10} \nabla^2 E_{ij})_0 \qquad (30)$$

where E_{ij} is the ambient rate of strain, and the subscript 0 indicates evaluation at the center of the test sphere. If the suspension is extremely dilute, the test sphere is not affected by other particles and $(E_{ij})_0 = \langle e_{ij} \rangle$, leading to the well-known expression due to Einstein for the effective viscosity of a dilute suspension of identical solid spheres to $O(c)$,

$$\mu^* = 1 + \frac{5}{2} c \qquad (31)$$

To calculate μ^* to $O(c^2)$, we must obtain S_{ij} to $O(c)$. Once again it would appear reasonable to suppose that, since the probability of a second sphere being within an $O(1)$ distance of the test sphere is itself $O(c)$, the required correction could be obtained by considering the influence of just that second sphere and then averaging over all its possible locations. Unfortunately, as in the previous problems, the second sphere induces an $O(R^{-2})$ disturbance velocity, or an $O(R^{-3})$ disturbance strain rate, which, according to eq. (30), gives rise to a non-absolutely convergent integral for the conditionally averaged stresslet at the test sphere.

To apply our effective continuum method we replace the suspension surrounding the test sphere with a concentric layer of pure fluid for $1 \leq r \leq 2$, and set the viscosity of the effective continuum equal to $1 + 5c/2$ for $r \geq 2$. As was done in the first example, we may solve this problem in two steps, the answers to which may then be superposed owing to linearity. Specifically, we first imagine that the test sphere is completely surrounded by fluid of viscosity $1 + 5c/2$ which increases the stresslet associated with an isolated sphere, and thereby the $O(c)$ term in eq. (31) by

(1+5c/2). The resulting expression for the effective viscosity is therefore

$$\mu^* = 1 + \frac{5}{2}c + \frac{25}{4}c^2 + O(c^3) \qquad (32)$$

from which we must subtract the effects due to the presence of the particle exclusion layer. To this end, we imagine that the test sphere is surrounded by clear fluid of unit viscosity everywhere except within the concentric shell $1 \leq r \leq 2$, where the effective viscosity is set equal to $1 + 5c/2$. This problem may now be solved by realizing that this layer of effective continuum may be replaced with a suspension containing a volume fraction c of tiny spheres, and then integrating over the influence of the latter on the stresslet at the test sphere. Since the region is finite, no convergence difficulties are encountered, and the combined approximate correction to the effective viscosity of the suspension is

$$\mu^* = 1 + \frac{5}{2}c + 4.02\,c \qquad (33)$$

Again, in order to obtain the exact $O(c^2)$ coefficient, it is necessary to account for all the other contributions to S_{ij} arising from the other terms of the exact two particle solution and then integrate the now well-behaved remaining corrections over the configurational ensemble. As shown by Batchelor and Green[9], such a calculation yields

$$\mu^* = 1 + 5c/2 + 5.01\,c^2 + O(c^3) \qquad (34)$$

using the more accurate value of the coefficient of the $O(c^2)$ term[3].

Through the solution of the three problems presented in this paper, we have seen then that the non-absolutely convergent and divergent integrals that arise upon taking conditional averages of certain relevant quantities can be resolved in a direct and physically consistent way. The advantage of this method over those presented earlier in the literature is that the source of the initial convergence problems is clearly shown to be due to neglecting the fact that the composite acts as an effective continuum far away from the test particle. In addition, in the absence of an exact two sphere solution, this method provides a means for estimating transport parameters in two-phase materials. Specifically, since in the approximate systems, where the suspension surrounding the test sphere is replaced by layers of effective continuum, more reflections are taken into account than merely the one corresponding to the lowest order interactions which are responsible for the convergence difficulties, the resulting values for the effective properties thereby derived would be expected to provide acceptable approximations to the exact ones. Still, although eqs. (15) and (25) yield predictions which are in close agreement with the corresponding exact values, eqs. (16) and (26), the approximation to the effective viscosity, eq. (32) is not as good, thereby suggesting that some care should be taken in the application of this method.

ACKNOWLEDGEMENT

This work was supported in part by the Department of Energy, under grant DE 13328.

REFERENCES

1. D. J. Jeffrey, Proc. R. Soc. Lond. A335, 355 (1973).
2. G. K. Batchelor, J. Fluid Mech. 52, 245 (1972).
3. H. S. Chen and A. Acrivos, Int. J. Solids Structures 14, 349 (1978).
4. R. W. O'Brien, J. Fluid Mech. 91, 17 (1979).
5. E. J. Hinch, J. Fluid Mech. 83, 695 (1977).
6. E. Y. Chang, B. S. Yendler, and A. Acrivos, to appear in the proceedings of the SIAM workshop on multiphase flow (1986).
7. Z. Hashin, J. Composite Materials 2, 284 (1968).
8. R. Landauer, J. Appl. Phys. 23, 779 (1952).
9. G. K. Batchelor and J. T. Green, J. Fluid Mech. 56, 401 (1972).
10. J. M. Rallison, J. Fluid Mech. 88, 529 (1978).

III. Chemistry

PROGRESSIVE CHEMICAL MODIFICATION OF CLASTIC SEDIMENTS WITH BURIAL

C. D. Curtis
Department of Geology, University of Sheffield
Sheffield S3 7HF, England

ABSRACT

The porosity of clastic sediments at deposition varies very approximately between about 45% (sands) and 85% (muds). With burial, consolidation takes place as pore water is progressively eliminated. It would be misleading, however, to attribute alterations in sediment bulk properties to physical processes alone. Very significant mineralogical changes occur and these start soon after burial, especially in mudrocks. Striking heterogeneities such as thin, laterally continuous cemented horizons or discrete concretions are commonly introduced. These shallow burial processes are predominently the result of microbial actvity. Thermodynamically unstable mixtures of organic matter and various oxidants (dissolved oxygen, sulphate, nitrate, particulate Fe(III) and Mn(IV)) provide both substrate and energy source for a variety of different microbial ecosystems. Mineralogical consequences include both leaching and the precipitation of carbonate, sulphide, phosphate and silica cements. The type and extent of mineral modification depends strongly on depositional environment variables such as rate of sedimentation and water composition.
At greater depths, large scale modification of detrital clay minerals (particularly the smectite-I/S-illite transformation) takes place. Recent work of various kinds, however, has demonstrated that these changes may not be solid state transformations: clay mineral dissolution, transport and precipitation occur much more widely than was formerly supposed. In sandstones, authigenic precipitation of clay minerals from pore solution is much more obvious. Systematic patterns of precipitation, alteration and replacement have been documented in many sedimentary basins. Porosity and permeability are reduced by cementation and, sometimes, enhanced by mineral dissolution. Whereas the general nature of these chemical reactions is fairly well understood, it is not yet possible to predict with certainty the scale or distribution of mineralogical consequences. Much debate, for example, surrounds the mechanism of porosity enhancement in sandstones. More information is needed about amounts and rates of porewater migration at different stages of compaction and the mobility of chemical solutes in the deep subsurface.

What is certain is that almost all clastic sediments encountered during deep drilling will have been modified very substantially by chemical processes during burial. Textural characteristics such as grain size and shape, fabric and packing will have been altered in consequence.

INTRODUCTION

This paper is concerned with diagenetic alteration of clastic sediments. The term diagenesis is here used to cover chemical modification of sediments from the point of deposition until buried to depths where temperatures of the order $200°C$ are encountered. Higher temperature transformations fall into the realm of metamorphism although, of course, the boundary between diagenesis and metamorphism is gradational. Organic geochemists use the term catagenesis to describe diagenetic processes taking place at temperatures greater than $50°C$ and this distinction is meaningful in that microbial processes tend to dominate alteration below this temperature but not above it.

Diagenetic processes may radically modify the porosity and permeability of sediments at any time throughout their burial history. This modification occurs on all scales from hand specimen to sedimentary basin with major consequences for fluid storage and migration. Both reduction and enhancement of porosity have been described in detail for numerous geological formations[1,2].

The driving force for diagenetic alteration is chemical instability. The various mineral constituents of detrital sediments at deposition are rarely, if ever, in chemical equilibrium with each other or with the solutes present in porewater. Diagenesis amounts to a sequence of reactions in which the least stable components react first, others subsequently. Pore waters are progressively eliminated and redistributed through the sediment column by physical compaction: chemical reactants are thus transported from one location to another and it is unsafe to assume closed-system alteration.

STARTING MATERIALS

Unconsolidated materials eroded from the land surface are the main constituents of sediments at deposition. They are augmented to greater or lesser extent by residues of organisms living in the water column - principally organic matter, carbonates and silica.

Most of the land surface is covered by soil profiles and it is these rather than bedrocks which are eroded. Soils can be considered to consist of three broad categories of material – coarse grained mineral matter, fine grained mineral matter and degraded organic matter. The first of these is derived essentially by physical breakdown of parent bedrock and its composition and stability will obviously reflect the latter. The fine fraction consists of phyllosilicate clay minerals together with both crystalline and amorphous hydrated oxides of Fe(III), Al and Si. Which clays and which hydrated oxides are present in any particular profile will depend on bedrock composition, topography and, especially, climate. At high latitudes physical processes dominate over chemical and soils consist mostly of bedrock minerals. At low latitudes with humid climate, chemical weathering is very much more important and soils consist largely of neoformed minerals – clays and hydrated oxides. Gradusov[3] has compiled maps of the global distribution of soil minerals which amply demonstrate these links. The distribution of clay minerals in sediments of present-day oceans[4] also reflects global climatic zones attesting to direct derivation from soils.

There are relatively few clay minerals stable in soil environments, kaolinite and smectite are the two most important.

$$\text{kaolinite} \quad Al_2[Si_2O_5](OH)_4$$

Soil smectite is related to muscovite in both structure

$$\text{muscovite} \quad K_2Al_4[Si_6Al_2O_{20}](OH)_4$$

and composition: it is basically a dioctahedral mica. The important reversible hydration and ion-exchange properties stem from a much lower interlayer charge which, in turn, results from much lower Al-for-Si tetrahedral substitution. Most soil smectite probably lies somewhere within the beidellite-nontronite compositional range:

$$\text{beidellite} \quad Na_{0.7}Al_4[Si_{7.3}Al_{0.7}O_{20}](OH)_4$$

$$\text{nontronite} \quad Na_{0.7}Fe_4[Si_{7.3}Al_{0.7}O_{20}](OH)_4$$

with Fe predominently Fe(III)

Where chemical weathering is significant but incomplete, a range of other clay minerals is found: hydromicas, vermiculite, illite, chlorite and mixed-layer minerals. In the main these are not truly stable but represent partial alteration of bedrock phyllosilicates. All but chlorite have structures and compositions somewhere intermediate between mica and smectite.

At deposition, therefore, clastic sediments contain coarse mineral grains whose composition and stability reflect the environment of formation of the original bedrock (obviously highly variable). The fine mineral fraction consists of clays and hydroxides formed in the hydrous and oxidizing soil environment. Finally, in fine-grained sediments, there may be a significant organic matter component.

REDOX REACTIONS IN THE UPPERMOST LAYERS OF BURIED SEDIMENTS: THE MICROBIAL ZONES (Ox, SR and Me)

When soils are eroded, transported and sedimented, separation of coarse (sands) from fine (muds) fractions normally takes place. The resulting sediment also contains pore water; upwards of 40% by volume in the former, 75% in the latter. Significant quantities of salts are trapped in marine and saline lake sediments.

Muddy sediments are chemically unstable since organic matter is a reducing agent and Fe(III) an oxidant. Oxygen and sulphate dissolved in porewater are two more oxidants in the system. Reactions between organic matter and these oxidants have been studied extensively in recent sediments. Within the first few tens of metres of sediment beneath oxic marine waters, a distinctive ecological succession is almost invariably found. Investigations of natural occurrences, laboratory experimental work and interpretations are reviewed in Claypool and Kaplan[5]. The first zone of this stratiform succession is characterised by aerobic respiration as principal metabolic process:

$$CH_2O + O_2 = H_2O + CO_2 \qquad (Ox)$$

Its lower limit is defined by downward diffusion of molecular oxygen in undisturbed sediments (usually no more than a very few cm). Bioturbation and water flushing by benthic organisms deepen the oxic zone. Various numbering systems have been proposed for these ecological zones which are just the first in a lengthier sequence of depth-related diagenetic zones (Curtis[6,7]). The simplest and most logical scheme is that recently introduced by Coleman[8]. Zones are denoted by abreviations according to the principal metabolic (or redox) reaction operating within them. Both zone and reaction can be referred to in this way and this is, in itself, useful.

Relatively soon after oxygen depletion, sulphate reduction becomes the dominant metabolic pathway; seawater being rich in dissolved SO_4^{2-} (0.028M). Sulphate-reducing bacteria thrive, probably in association with other anaerobic bacteria which convert organic substrates into more readily metabolisable form. Hydrogen sulphide and carbon dioxide are the two end-products:

$$2CH_2O + SO_4^{2-} = HS^- + HCO_3^- + H_2O + CO_2 \quad \text{(SR)}$$

Sulphate reduction is limited mainly by availability of SO_4^{2-} (diffusing down from overlying depositional water) and suitable substrate organic matter. The zone of SR dominance tends to be about a metre thick, increasing to several m. in organic-poor sediments.

Beneath the lower limit of sulphate penetration, methane is found. The mechanisms of formation and environmental constraints were the focal point of Claypool and Kaplan's paper. Microbial pathways were discussed and it was concluded that bacterial reduction of CO_2 was most important. Other interpretations are possible but there is little doubt that complex assemblages of microorganisms must be involved and relatively little is known about them. Overall, suitable organic substrate is broken down to produce methane and some more oxidised product; probably CO_2:

$$2CH_2O = CH_4 + CO_2 \quad \text{(Me)}$$

One very important contribution made by Kaplan's group (Presley and Kaplan[9], Presley et al.[10], Nissenbaum et al.[11], Claypool et al.[12], amongst others) was to follow fractionation of stable carbon and sulphur isotopes in these reactions. An elegant study of freshwater microbial isotope fractionation had been earlier published by Oana and Deevey[13]. The relevance to studies of inorganic geochemistry is in the preservation of these fractionations in carbonate and sulphide minerals.

Lack of fractionation within the Ox and SR Zones produces bicarbonate from organic matter which is very different from normal marine bicarbonate (the former being enriched in ^{12}C by some 2.5% relative to the latter). Diagenetic carbonate minerals incorpoating carbon from these sources is thus readily identifiable. The same is true of bicarbonate derived from microbial methanogenesis (distinctively "heavy" bicarbonate).

The other oxidant mentioned above was Fe(III). Recently buried sediments often include carbonates and sulphides which contain both Fe^{2+} and Mn^{2+} (Suess[14]. Obviously mineral oxidants must have been involved in organic matter degradation and the reduced products incorporated together with CO_3^{2-} and S^{2-} from the Ox, SR and Me reactions.

Oxides and hydroxides of Mn(IV) and Fe(III) are both present in soils and can therefore act as oxidants:

$$CH_2O + 2MnO_2 + H_2O = 2Mn^{2+} + HCO_3^- + 3OH^- \quad \text{(MnR)}$$

$$CH_2O + 2Fe_2O_3 + 3H_2O = 4Fe^{2+} + HCO_3^- + 7OH^- \quad (FeR)$$

Both reactions increase alkalinity (Curtis [6,7], Coleman[8]) and would thus be expected to encourage carbonate precipitation. Development of very high pH environments due to extensive FeR or MnR, however, would destabilise silicates, including quartz. This is because most aluminosilicates have solubility minima at near-neutral pH values: quartz solubility increases to high pH only (8 plus).

Pelagic sediments accumulating very slowly develop up to six distinct diagenetic depth zones just below the sediment water interface (Froelich et al.[15]). Beneath the aerobic respiration zone, nitrate is reduced (NR). Then come distinct MnR and FeR zones prior to complete anoxia and sulphate reduction.

In more rapidly accumulating sediments with abundant organic matter, separate MnR and FeR zones cannot be recognized: metal reduction and microbial degradation (SR and Me) occur together. High sulphide and/or bicarbonate activities and rapid rates of mineral precipitation maintain pore water metal activities at very low levels. The rate-determining step is probably reduction of the solid oxidants. It would appear that Mn(IV) is more easily reduced than Fe(III). It is certainly true that the latter survives to km depths in shelf clastic sequences (Foscolos and Powell[16]). Hydrated Fe(III) oxides from soil profiles have large surface areas and are probably reduced and incorporated into diagenetic Fe(II) minerals soon after sediment burial. Somewhat less reactive Fe(III) is present as an interlayer complex in some soil smectites. That in the octahedral layer of nontronite-beidellite clays can only be released by major structural modification. It can be argued that, with increasing depth of burial, the less reactive Fe(III) compounds become progressively more important reactants.

SHALLOW BURIAL MINERAL REACTIONS

All the reactions described above lower the organic content of shales and, incidentally, their potential for generating hydrocarbons. Solute products of these reactions supersaturate pore water with respect to several minerals. These then precipitate, displacing pore water and may cement the enclosing sediment. Each zone stabilises its own distinct assemblage:

ZONE Ox; oxidation by molecular oxygen

Organic matter is degraded by respirative bacteria and benthic fauna. The lower limit of the zone is probably determined by the water circulating activity of burrowers (usually a few cm). This mixing will prevent high solute concentrations building up hence significant mineral authigenesis is unlikely except for iron and manganese oxides in sites of low sedimentation rate. Fe(III) should be the stable valence state and reduction will not occur except in transient or localised microenvironments.

ZONE SR; oxidation by sulphate and Fe(III)

The following equation represents the sum of SR and FeR when the amount of iron reduced is exactly balanced by sulphide produced:

$$15CH_2O + 2Fe_2O_3 + 8SO_4^{2-} = 4FeS_2 + H_2O + 15HCO_3^- + OH^-$$

Pyrite precipitation is thus accompanied by a very large increase in pore water bicarbonate activity without lowering pH. Fe-poor carbonates (FeS_2 is highly insoluble) are very likely to form alongside pyrite (calcite, dolomite - partly at the expense of metastable primary carbonates such as aragonite). Phosphate released by organic degradation will precipitate (francolite; a complex Ca-phosphate mineral). Manganese reduction (MnR) will also occur in the SR Zone with Mn incorporated into both carbonates and sulphides.

ZONE Me; Methanogenesis and oxidation by Fe(III)

The combination of microbial methanogenesis with Fe(III) reduction is extremely conducive to carbonate precipitation; but this time of iron-rich carbonates (siderite, ankerite, ferroan dolomite) since no sulphide is available to remove Fe^{2+}:

$$7CH_2O + 2Fe_2O_3 = 3CH_4 + 4FeCO_3 + H_2O$$

In this combination of Me and FeR, the two contributions are just sufficient to precipitate HCO_3^- and Fe^{2+} quantitatively. An excess of iron reduction would lead to alkaline pore waters and, perhaps, precipitation of Fe(II)-silicates such as berthierine. Conversely, depletion of the iron supply would create less alkaline (even acid) environments wherein phosphate minerals are stabilised relative to carbonates (Nathan & Sass[17]).

A common feature of clastic sedimentary sequences is the occurrence of concretions. These are localised concentrations of carbonate, sulphide or phosphate cement within (usually) less well cemented rocks. Detailed investigations of structure, chemical composition and stable isotope ratios show them to be the direct result of the processes described above[18,19]. These mineral segregations effectively produce profound heterogeneities of strength (cementation) and composition which certainly affect drilling in the former case and, in all likelihood, downhole logging techniques in the latter.

ORGANIC MATTER DIAGENESIS AND CATAGENESIS

The organic content of muds is lowered and extensively modified by microbial processes. Work in this area has been summarized by Hunt[20]. Biopolymers (carbohydrates, proteins etc.) are broken down to low molecular weight biomonomers (sugars, amino acids) and some are utilised as substrate by bacteria, being oxidized in the process. Those surviving are chemically reactive and tend to recombine in condensation reactions to produce geopolymers - protokerogen. Whereas the original biopolymers were present mostly as discrete particles (detrital organic residues), geopolymers are likely to have formed in intimate association with the fine grained and high surface area clay fraction. The physical properties of the resulting mudrock will be affected by organic coatings via their influence on inter-particle forces. It may well be that the nature (hydrophobic, hydrophyllic) and amount of organic matter are important controls on bulk mudrock properties.

As temperatures within the sediment rise with burial, organic matter starts to break down via thermocatalytic cracking. Early products are H_2O and CO_2 (Tissot et al.[21], Laplante[22]). The latter also can become incorporated into diagenetic carbonate cements (Irwin et al.[23]). Excess CO_2 and organic acid production, however, generate acid environments within which carbonate and silicate minerals tend to dissolve. Many oilfield brines have been found to contain high concentrations of acetate (Carrothers and Kharaka[24]) and dissolution porosity is commonplace[1,2]. Considerable controversy remains, however, as to the quantitative significance of organic acid and CO_2 generated porosity.

Thermocatalytic processes thus must be anticipated to influence sediment properties by modifying organic matter in intimate association with clay minerals and, possibly, by causing dissolution of cements and destabilization of framework grains.

DIAGENETIC MODIFICATION OF SILICATE MINERALS

Drever[25] showed that relatively little modification of detrital clays takes place during early diagenesis within the microbial zones. Such as there is amounts mostly to ion exchange. Organic-derived NH_4^+ may be incorporated in interlayer sites.

The petroleum industry enabled systematic studies of deeper clay diagenesis by providing core and cuttings from deep boreholes through sediment sequences with relatively simple burial histories. An early interest centred on the possibility that clay mineral dehydration reactions could generate high fluid pressures and play some part in assisting primary migration of hydrocarbons from sourcerocks (Powers[26], Burst[27]). More important here, however, was the fact that these studies confirmed the general pattern of clay mineral depth distribution described earlier by Burst[28] in U.S. Gulf Coast Tertiary mudrocks.

Using powder XRD methods, Burst had shown that true (fully expandable) smectite was present in mudrocks buried to 1km but no deeper. An intermediate depth zone (1-3km) was characterised by progressive diminution of swelling properties after which (to 4.5km), only illite persisted. The latter was seen to have a generally asymmetric 1.0nm reflection. The two deeper zones also seemed to contain increasing amounts of chlorite which, with greater depth, demonstrated improved crystallinity. A number of independent studies came to essentially similar conclusions (reviewed by Dunoyer de Segonzac[29]).

Without neglecting the possibility of systematic changes in detrital sediment composition, these trends were interpreted as indicating the progressive alteration of smectite through intermediate, interstratified illite/smectite (I/S) clays, to illite. The implication was that transformation had occurred with the basic phyllosilicate structural framework retained. Presumably the improving chlorite crystallinity could be attributed to the beneficial effect of reducing conditions: Fe(III) reverting to Fe(II) which would have been the prevalent oxidation state in the original, unweathered chlorites.

This general pattern of burial diagenesis in mudrocks was expanded upon in 1976 by a series of papers from Hower and his associates (notably Hower et al.[30], Aronsen and Hower[31]). Shale cuttings were obtained from Oligocene-Miocene sediments of the U.S. Gulf Coast. They were carefully disaggregated, separated into several size fractions and each was subjected to detailed examination by X-ray powder diffraction, chemical analysis and mass spectrometry. The intermediate depth zone (Burst[27]) showed major alteration with I/S clays (the most abundant clay mineral) undergoing progressive conversion from less than 20% to about 80% illite layers, as judged from orientated powder diffraction patterns. The coarse fraction lost calcite, potassium feldspar (but not albite) and kaolinite over the same depyh interval with an increase in chlorite content. Losses from the coarse fraction could be equated to some extent (not carbonates) with fine-fraction clay mineral transformations:

$$Al^{3+} + K^+ + smectite = Si^{4+} + illite$$

Potassium was thought to be supplied by detrital feldspar and mica breakdown. This being the case, radiogenic argon would be lost in the transformation since very little would be incorporated with potassium into the more illitic clay. This pattern was documented by Aronson and Hower[31]; coarser fractions (richer in detrital feldspar) giving higher $^{39}Ar/K$ ratios than finer and both fractions showing lower values at greater depths (apparently "younging" downwards). The consistency between these different types of evidence and similar findings by other workers led to rather general acceptance of the transformation mechanism implied by the equation given above.

Boles and Franks[32] studied diagenetic alteration in both shales and sandstones; they were concerned with possible links between the two in mixed sequences. They also reviewed earlier published data and estimated the compositions of "end-member" illite and smectite in typical Gulf Coast I/S clays. Such data allow better approximations to clay transformation equations although difficult assumptions must still be made. Boles and Franks[32] suggested two instructive cases: Al mobile (all aluminium released by feldspar degradation migrating via pore solutions and being incorporated in clay transformation thereby generating an overall increase in clay content) and Al immobile (I/S transformation by potassium addition and silica loss: molar replacement of smectite by illite).

a) Al mobile: $4.5K^+ + 8Al^{3+} + smectite = illite + Na^+ + 2Ca^{2+} + 2.5Fe^{3+} + 2Mg^{2+} + 3Si^{4+} + 10H_2O$

b) Al immobile: $3.93K^+ + 1.57smectite = illite + 1.57Na^+ + 3.14Ca^{2+} + 4.28Mg^{2+} + 4.78Fe^{3+} + 24.66Si^{4+} + 57O^{2-} + 11OH^- + 15.7H_2O$

It is quite clear that either (or any intermediate) formulation releases a range of solutes to pore water which could precipitate as authigenic cements within either shales or nearby sandstones. Boles and Franks[32] suggested that likely reactions were precipitation of authigenic quartz and albite with conversion of calcite to ankerite and kaolinite to chlorite. Problems of timing were recognised (chloritization seems to take place significantly deeper than the main phase of I/S transformation) but this basic pattern has been repeatedly described, even in very recent reports (Morton[33], Pollastro[34]). Geological environments with very high geothermal gradients such as the Salton Sea region (Muffler and White[35], McDowell and Elders[36]) have yielded similar mineralogical progressions on much condensed time and depth scales.

In summary, the principal diagenetic change that has been deduced from data based on a combination of chemical and X-ray methods is loss of true smectite followed by progressive transformation of I/S clays during which smectite layers are modified to illite. Some workers (Foscolos and Powell[16] for example) have questioned the transformation mechanism and suggested dissolution of the smectite component as an alternative. It is noteworthy that an attempt was also made to document progressive loss of such inherited soil components as hydrated iron hydroxides. Attempts to relate mudrock mineralogy back to typical precursor soil mineralogy are none too common.

In one sense the story of diagenesis in sandstones is quite different from that in mudrocks and, incidentally, much less difficult to follow. A combination of thin section optical petrography and X-ray powder diffraction is sometimes sufficient to demonstrate that neoformation is a very important origin mechanism. It can be seen that clays of uniform colour and optical properties coat framework grains or occupy pore space. Scanning electron micrographs reveal delicate and beautiful growth patterns that could not possibly have survived physical transport (Wilson and Pittman[37]). Based on information of this type, Hayes[38] was able to demonstrate that neoformed chlorite, as grain coatings, was a characteristic clay constituent (often the only one) in sandstones from a range of localities and geological histories. Precipitation from pore water solutes would appear to be the only sensible mechanistic explanation.

The sedimentological literature abounds with perfectly composed pictures of clays in reservoir sandstones. The story for chlorite fits equally well for neoformation of smectite, kaolinite and illite. Not infrequently the clay assemblage is, to all intents and purposes, monominerallic. Samples can be separated and studied by XRD which often yields beautifully sharp reflections. This contrasts with traces from many mudrocks.

There are, of course, many "dirty" sandsones which contain inherited clays little different from those in mudrocks. Amongst "cleaner" sands, however, the pattern of diagenesis is characteristically one of neoformation from aqueous solution rather than transformation of inherited clays.

Studies of mudrocks and sandsones in the same succession (Boles and Franks[32], McDowell and Elders[36]) do reveal similarities. Kaolinite does not survive to great depth whereas neoformed chlorite is characteristic of moderate to great depths. Authigenic illite generally postdates (often replaces) neoformed kaolinite. The trouble in mudrocks is distinguishing between inherited and neoformed fractions of the same mineral. In the absence of such distinction, the pattern of diagenesis in mudrocks will be blurred and very difficult to interpret.

Much more detailed information about the structure and composition of clay minerals has recently emerged from the application of a variety of transmission electron microscopic techniques[39-57]. This confirms the authigenic (i.e. formed directly from solution) nature of many sandstone clays but casts doubt on a gradual transformation of detrital clays (inherited from soils) from smectite through to illite and chlorite. It appears that a dissolution-precipitation mechanism is more likely.

No matter what the details of mechanism turn out to be, however, there can be no doubt that major changes in silicate mineralogy take place in the 1 to 5km burial depth range. Clay mineral cements precipitate in sandsones and obviously modify porosity, permeability, strength and many other interesting properties (such as resistivity). Precipitation of a continuous film of clay through connected sandstone pore space (particularly common in the case of chlorite where quartz overgrowths can be prevented) would obviously be relevant in this context. The newer electron microscope studies of mudrocks are starting to suggest similar textural modifications; again with potentially important implications for physical properties. There is ample evidence of depth-related silicate dissolution reactions in both sandstones and mudrocks. These must reduce strength and enhance porosity. The opposite effects are encountered when precipitation of quartz or feldspar occurs as interlocking overgrowths – again a common event.

CONCLUSION

Within the context of a symposium devoted to the physics and chemistry of porous media, this brief review has attempted to outline the framework within which sedimentologists and sediment geochemists currently view chemical alteration of sediments as they are buried to depths of several km. It should be clear that changes are major and, to some extent, predictable. It also appears that physical properties of sedimentary rocks relevant to the drilling industry must be modified to significant extent by diagenetic processes. Reviews such as this can never hope to be comprehensive. It is hoped, however, that within the literature cited here, useful pointers to future research may be found.

ACKNOWLEDGEMENTS

This review includes many observations based on the work of former (and present) PhD students at Sheffield. I am indebted to them as also to many colleagues. Financial support for work here has come variously from The Natural Environment Research Council, Exxon, Amoco and BP. All are gratefully acknowledged.

REFERENCES

1. Hayes, J. B. (1979). SEPM Spec. Pub. 26, 127-139.
2. Schmidt, V. and McDonald, D. A. (1979). SEPM Spec. Pub. 26, 175-207.
3. Gradusov, B. P. (1971). Dokl. Akad. Nauk. SSSR., 202, 1164-1167.
4. Griffin, J. J., Windom, H. and Goldberg, E. D. (1968). Deep Sea Research, 15, 433-459.
5. Claypool, G. E. and Kaplan, I. R. (1974). In "Natural Gases in Marine Sediments" (Kaplan, I. R., Ed.), Plenum, New York.
6. Curtis, C. D. (1977). Phil. Trans. R. Soc. London, A286, 353-372.
7. Curtis, C. D. (1980). J. Geol. Soc. London, 137, 189-194.
8. Coleman, M. L. (1985). Phil. Trans. R. Soc. London, A315, 39-56.
9. Presley, B. J. and Kaplan, I. R. (1968). Geochim. Cosmochim Acta, 32, 1037-1048.
10. Presley B. J., Goldhaber. M. B. and Kaplan, I. R. (1970). Init. rep. DSDP 5, 517-522.
11. Nissenbaum, A., Presley, B. J. and Kaplan, I. R. (1972). Geochim. Cosmochim. Acta, 36, 1007-1027.
12. Claypool G., Presley, B. J. and Kaplan, I. R. (1973). Init. Rep. DSDP 19, 879.
13. Oana, S. and Deevey, E. S. (1960). Am. J. Sci., 258-A, 253-272.
14. Suess, E. (1979). Geochim. Cosmochim. Acta, 43, 339-352.
15. Froelich, P. N., Klinkhammer, G. P., Benber, M. L., Luedtke, N. A., Heath, G. R. Cullen, D., Dauphin, P., Hammond, D., Hartman, B. and Maynard, V. (1979). Geochim. et Cosmochim. Acta, 43, 1075-1090.
16. Foscolos, A. E. and Powell, T. G. (1980). Can. Soc. Pet. Geol., Mem. 6, 153-172.
17. Nathan, Y. and Sass, E. (1981). Chem. Geol., 34, 103-111.
18. Curtis, C. D. and Coleman, M. L. (1986). SEPM Spec. Pub. 38, Ed. D. L. Gautier, pp23-33.
19. Curtis, C. D., Coleman, M. L. and Love, L. G. (1986). Geochim. et Cosmochim., in press.
20. Hunt, J. M. (1979). "Petroleum Geochemistry and

Geology", Freeman, San Francisco, 617pp.
21. Tissot, B., Durand, B., Espitalie, J. and Combaz, A. (1974). AAPG bull., 57, 499-506.
22. Laplante, R. E. (1974). AAPG Bull., 53, 1281-1289.
23. Irwin, H., Curtis, C. D. and Coleman, M. L. (1977). Nature, London, 269, 209-213.
24. Carothers, W. W. and Kharaka, Y. K. (1978). AAPG Bull., 62, 2441-2453.
25. Drever, J. I. (1971). J. Sed. Pet., 41, 951-961.
26. Powers, M. C. (1967). A.A.P.G. Bull., 51, 1240-1254.
27. Burst, J. F. (1969). A.A.P.G. Bull., 53, 73-93.
28. Burst, J. F. (1959). Proc. 6th Natl. Conf. Clays and Clay Minerals. Pergamon. 327-331.
29. Dunoyer de Segonzac, G. (1970). Sedimentology, 15, 281-346.
30. Hower, J., Eslinger, E. V., Hower, M.E. and Perry, E. A. (1976). Bull. Geol. Soc. Amer., 87, 725-737.
31. Aronsen, J. L. and Hower, J. (1976). Geol. Soc. Am. Bull., 87, 738-744.
32. Boles, J. R. and Franks S. G. (1979). J. Sed. Pet., 49, 55-70.
33. Morton, J. P. (1985). Geol. Soc. Am. Bull., 96, 114-122.
34. Pollastro, R. M. (1985). Clays, clay minerals, 33, 265-274.
35. Muffler, L. J. P. and White, D. E. (1969). Geol. Sos. Am. Bull., 80, 157-182.
36. McDowell, S. D. and Elders, W. A. (1980). Contributions Mineralogy, Petrology, 74, 293-310.
37. Wilson, M. D. and Pittman, E. D. (1977). J. Sed. Pet., 47, 3-31.
38. Hayes, J. B. (1970). Clays, Clay Minerals, 18, 285-306.
39. Nadeau, P. H. (1985). Clay Minerals, 20, 499-514.
40. Nadeau, P. H., Tait, J. M., McHardy, W. J. and Wilson, M. J. (1984a). Clay Minerals, 19, 67-76.
41. Nadeau, P. H., Wilson, M. J., McHardy, W. J. and Tait, J. M., (1984b). Science, 225, 923-925.
43. Nadeau, P. H., Wilson, M. J., McHardy, W. J. and Tait, J. M., (1984c). Clay minerals, 19, 757-769.
44. Nadeau, P. H., Wilson, M. J., McHardy, W. J. and Tait, J. M., (1985). Mineralogical Magazine, 49, 393-400.
45. Ireland, B. J. (1985). Unpublished PhD thesis, University of Sheffield.
46. Ireland, B. J., Curtis, C. D. and Whiteman, J. A. (1983). Sedimentology, 30, 769-786.
47. Curtis. C. D., Ireland, B. J., Whiteman, J. A., Mulvaney, R. and Whittle, C. K. (1984). Clay Minerals, 19, 471-481.
48. Curtis, C. D., Hughes, C. R., Whiteman, J. A. and Whittle, K. (1985). Mineralogical Magazine, 49,

375-386.
49. Whittle, C. K. (1985). Unpublished PhD thesis, University of Sheffield.
50. Ahn, J.H. and Peacor, D.R. (1985a). Clays, Clay Minerals, 33, 228-236.
51. Ahn, J.H. and Peacor, D.R. (1985b). Abs. 8th Int. Clay Conf., Denver, Colorado. p4.
52. Ahn, J.H. and Peacor, D.R. (1986a). Proc. 8th Int. Clay Conf., Denver, Colorado., in press.
54. Ahn, J.H. and Peacor, D.R. (1986b). Clays, Clay Minerals, in press.
55. Lee, J. H. and Peacor, D. R. (1983). Nature, 303, 608-609.
56. Lee, J. H., Peacor, D. R., Lewis, D. D. and Wintsch, R. P. (1984). Contributions Mineralogy, Petrology, 88, 372-385.
57. Lee, J. H., Ahn, J. H. and Peacor, D. R. (1985). J. Sed. Pet., 55, 532-540.

THE MAJOR ION CHEMISTRY OF SALINE BRINES IN SEDIMENTARY BASINS

Lynton S. Land
Dept. Geological Sciences
University of Texas, Austin, Tx. 78713

ABSTRACT

The salinity of saline brines in most sedimentary basins increases with depth, and commonly approaches halite saturation. Additionally, $CaCl_2$ becomes an increasingly abundant component with increasing depth (salinity or temperature). Both these observations can be explained by density stratification of fluids derived from evaporite dissolution, and increased rock-water interaction with increased depth (temperature).

INTRODUCTION

The water which characteristically accompanies the production of oil and/or gas from sedimentary basins ranges widely in total dissolved solids content and in ionic composition. Every sedimentary basin contains somewhat unique brines, but two generalizations apply to almost all basins, especially those containing appreciable thicknesses of Paleozoic rocks. First, the total dissolved solids content of water commonly exceeds seawater, and increases nearly monotonously with increasing depth[1]. In many cases halite saturation is approached. And second, NaCl and $CaCl_2$ comprise almost all of the dissolved solids[2], in contrast to seawater where NaCl and $MgSO_4$ dominate. Increased calcium content commonly correlates with increased salinity.

ORIGIN OF THE ANION (CHLORIDE) CONCENTRATION

Most sediments are deposited in marine environments, and therefore seawater is a useful reference solution against which pore fluids produced from sedimentary rocks (or found as fluid inclusions in minerals) can be compared. Although water less saline than seawater occurs in many basins, solutions more concentrated than seawater are much more common. Chloride is the dominant anion in both seawater and in almost all saline subsurface water. Seawater contains approximately 2.7 g/L dissolved sulfate, but sulfate is easily reduced bacterially at temperatures below about 80°C, and thermally at higher temperatures. Sulfate-dominated solutions are therefore very rare in the subsurface. Bicarbonate concentrations are limited by carbon dioxide partial pressures and the solubility of calcite. Bromide and organic acid anions, principally acetate, are the only anions other than chloride which commonly reach, but rarely exceed, concentrations of 1 or 2 g/L.

Several theories have been advanced to explain how chloride might be concentrated in the subsurface beyond the value typical of modern seawater (19.4 g/L). Evaporation accompanying the formation of natural gas is not favored because of the vast quantities of gas required to remove sufficient water vapor to generate solutions containing up to about 210 g Cl/L. Gravitational settling of ions is offset by other forces, especially at elevated temperatures in the subsurface[4].

The only currently favored hypothesis which seeks to explain how the chloride derived from connate seawater might be concentrated in the subsurface is the concept of shale membrane filtration (reverse osmosis, or salt sieving)[5]. When a pressure gradient is imposed across a shale bed containing clay minerals, cations (especially small ones of low ionic charge) are more easily expulsed than Cl⁻, which tends to remain behind. In this way, brines progessively more saline and more enriched in calcium relative to sodium (and other large, multivalent cations, and heavy isotopes) might be produced on the high pressure side of the shale membrane. Although this theory still claims adherants, several rather fatal objections have been raised. Some authors have doubted that sufficient pressure gradients can be generated in nature[6], the chemistry of brines in shale-rich sections does not seem to conform to the expected pattern (Figure

1), and very wide ranges in formation water chemistry are observed in single formations consisting of similar types of sediments deposited in similar depositional settings (Figure 1). Most authors have recognized that the near halite-saturated, Ca-rich brines which are typical of most sedimentary basins, must in some way be related to evaporites, and not to the subsurface concentration of dissolved solids in connate seawater.

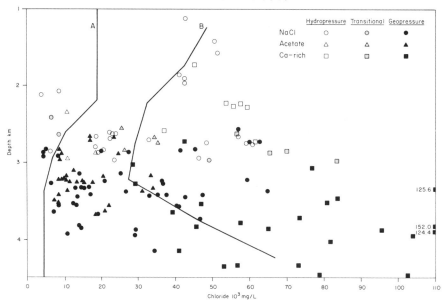

Fig. 1. Chloride content versus depth for formation water, Frio Formation (Oligocene), Texas Gulf Coast. The least saline brines are most common near the top of "hard" geopressure (line "B" is a moving average), where the saline residium should be concentrated if membrane filtration were a dominant process. A wide range of brine salinity and chemistry characterizes this deltaic sequence, and Ca-rich brines (derived from underlying strata) become more saline in the deepest reservoirs sampled. Line "A" depicts the salinity change which would occur if connate seawater were diluted by water released from the transformation of smectite to illite[3].

Why should the salinity of brines increase with increasing depth in so many basins? The answer to this question is probably that fluid flow is sufficiently uninhibited over geologic time that sedimentary basins simply become density stratified. The density of water is controlled by three variables: temperature, salinity, and pressure. Pressure is the least important of the three, and salinity the most important. For example, an increase in temperature from 25 to 200°C (a completely arbitrary "basement" for sedimentary basins) causes a decrease in density of about 0.075 g/cm^3. An increase in salinity from no salt (meteoric, or "fresh" water) to NaCl-saturation corresponds to an increase in density of between 0.15 and 0.2 g/cm^3, depending on temperature[7]. Therefore, in sedimentary basins, the effect of temperature (causing water density to decrease with increasing depth) is easily offset by salinity increases. Many sedimentary basins appear to be density stratified. Saline fluids derived from the dissolution of evaporite minerals deposited in the sedimentary rock section, and connate saline pore fluids, ultimately seek gravitational stability in exactly the same way as ocean water responds to perturbations of density (by evaporation, temperature changes, or the freezing out of ice) and seeks gravitational equilibrium. Density stratification can be "upset", especially in young or deforming basins where other hydrologic forces such as gravitational or tectonic compaction, forced flow, or hydrothermal convection are active. But in the absence of such forces, "thermohaline" circulation must result in density stratification, which in sedimentary basins is most obviously expressed as an increase in salinity with depth.

What is the source of the chloride - the burial of brines which formed at the Earth's surface as a result of precipitation of evaporite minerals from seawater, or subsurface dissolution of the evaporite minerals themselves? Several geologic observations suggest that evaporite minerals (principally halite) are destroyed by dissolution in the subsurface to give rise to saline brines. Ancient evaporites are known to become progressively rarer in progressively older rocks back through geologic time. Yet evaporites have always been deposited as part of the rock record. It has been estimated that the present mass of evaporites in the Earth's crust has been recycled (dissolved and reprecipitated) 15 times in the last 3 billion years[8]. This compares to a recycling rate three times faster than that of shales and 1.5 times faster than that of limestones. We know that the "residence time" of chloride in the world's ocean is on the order of 200,000,000 years[8], a smaller number by at least an order of magnitude than the time span of the sedimentary rock record. These considerations indicate significant recycling of chloride through Earth's crust. How is this recycling accomplished? Evaporites can, of course, be uplifted and eroded (dissolved) like other rocks, but then why are they destroyed and re-formed faster than limestones? An alternative is for evaporites to dissolve in the subsurface. The only evidence of ancient evaporite strata in some basins may be the saline brines themselves, and rock types like stratiform "collapse breccias", easily overlooked, or misidentified as tectonic or karst breccias. When the depositional facies of most basins, especially those containing abundant limestones, are examined in light of modern knowledge of the sedimentary structures associated with environments of evaporite deposition, "occult" evaporites are nearly always evidenced.

Another line of evidence how rapidly subsurface evaporite solution may operate, and how dynamic the flux of chloride in sedimentary basins may be, comes from the study of the chloride content of rivers, lakes and potable groundwater. Hydrogeochemists have experienced difficulty in explaining the source of chloride in surficial (potable) water[9]. Only small amounts of chloride can be due to aerosols derived directly from the ocean, and too little chloride is present in the minerals in rocks to yield the observed concentrations by mineral weathering. One potential solution to this dilemma is to postulate that the chloride which is discharged by rivers to the oceans is derived from the subsurface dissolution of evaporites, and enters the rivers and the surficial hydrologic system by vertical discharge from sedimentary basins (Figure 2).

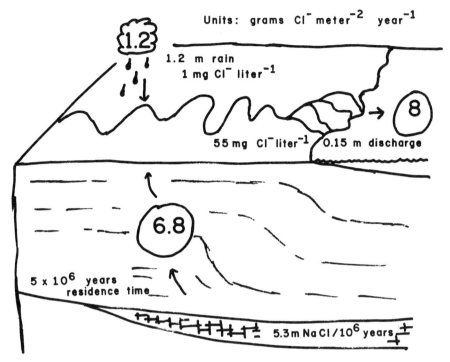

Fig. 2. Chloride budget for typical Texas rivers having their headwaters in the coastal plain. Rivers deliver about 8 grams of Cl^- per square meter of drainage area per year to the Gulf of Mexico, yet only 1.2 grams can be due to aerosols. Postulated discharge of 6.8 g from the subsurface requires the dissolution of 5.3 meters of halite per million years and yields a residence time of 5 million years for the formation water in the rock column.

Considerations of the dynamics of evaporite deposition throughout geologic time seem to suggest that significant amounts of chloride in saline brines in sedimentary basins derive from the dissolution of bedded halite deposited in the sedimentary section. Barring forced flow or hydrothermal convection, basins eventually become stratified in their salinity distribution until physical and chemical compaction force the saline pore fluids to the Earth's surface, until they are displaced (washed out) by meteoric recharge, or entombed as fluid inclusions in minerals.

ORIGIN OF THE CATION (ESPECIALLY CALCIUM) CONCENTRATION

It has long been recognized that sodium and calcium are the dominant cations in almost all brines found in sedimentary basins. The reason for this is simply that mineral equilibria do not permit other cations to achieve high concentrations relative to sodium and calcium. Figure 3 shows the relation between the molar calcium/magnesium ratio of a solution in equilibrium with the common sedimentary minerals calcite and dolomite as a function of temperature. Although complexities exist in relating molar ratios to activity ratios (assumed identical here), and in non-ideal behavior of the solids, particularly dolomite, it is uniformly agreed that equilbirium between these two minerals demands that calcium must exceed magnesium at sedimentary basin temperatures. In fact, the calcium/magnesium ratio of a great many basins can be used as a rude "geothermometer"[11]. Likewise (Figure 4), sodium must exceed potassium based on equilibria between the common minerals albite (Na-feldspar) and K-feldspar[12]. Figure 4 also shows why sodium must dominate over calcium. Plagioclase feldspar is the most abundant mineral in the Earth's crust, and albite (Na-plagioclase) is the most stable (least soluble) feldspar at the relatively low temperatures characteristic of sedimentary basins. Therefore, at sedimentary basinal temperatures, sodium-rich solutions must prevail.

Mineral equilibria among the most common minerals which characterize sediments at sedimentary basinal temperatures dictate that sodium must dominate both calcium and potassium (Figure 4), and calcium must dominate magnesium (Figure 3). Other common cations such as strontium and barium, which can reach concentrations of several grams/L, are prohibited from reaching higher values by relatively insoluble minerals like barite ($BaSO_4$), celestite ($SrSO_4$), witherite ($BaCO_3$), and strontianite ($SrCO_3$).

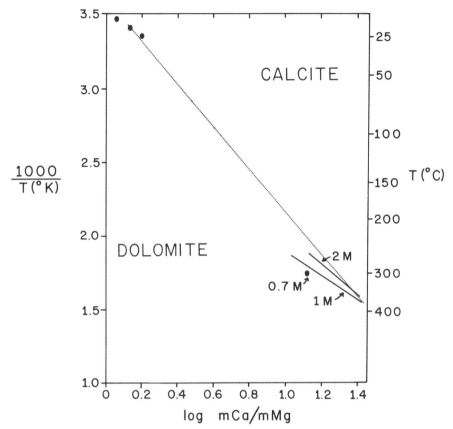

Fig. 3. Log of the molar calcium/magnesium ratio of a solution in equilbrium with calcite plus dolomite[10]. Calcium must dominate magnesium at all basinal temperatures in solutions at equilibrium with these two common minerals.

During burial, plagioclase feldspars (containing variable amounts of Ca) are replaced by albite and K-feldspars commonly dissolve[13] (Figure 5). In sandstones these processes commonly produce megascopic (reservoir) porosity at depths corresponding to temperatures between about 80 and 120°C[15], but they take place in shales as well[14]. Feldspars contain abundant strontium, barium, and many other trace elements (including lead and zinc), and the origin of these elements in saline brines is almost certainly related, at least in part, to extensive feldspar-water reaction.

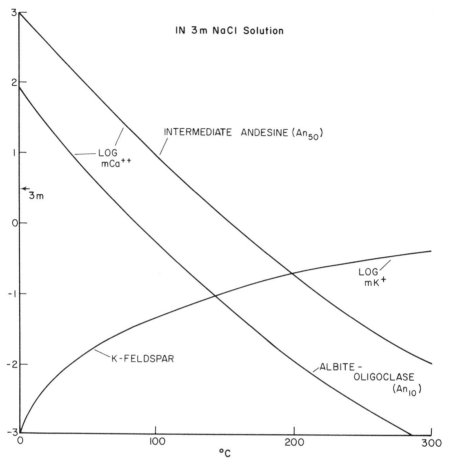

Fig. 4. Log molar concentration of potassium (lower curve) in equilibrium with K-feldspar in 3 m NaCl solution as a function of temperature. Upper curves are log molar concentration of calcium in equilibrium with two kinds of typical detrital plagioclase in 3 m NaCl solution. At temperatures above about 80°C, where kinetic constraints on feldspar reactions are overcome, sodium must dominate both calcium and potassium in subsurface water if feldspar equilibria prevail[12].

An alternative to the hypothesis that subsurface brines owe their chemistry primarily to halite dissolution and subsequent rock-water interaction in the subsurface is that they derive from brines from which the evaporites precipitated in the first place[16]. The evap-

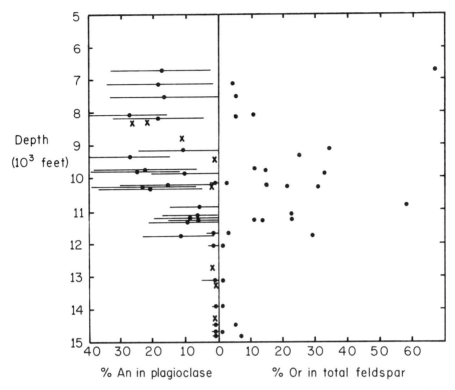

Fig. 5. Feldspar compositions, Frio Formation (Oligocene), Texas Gulf Coast. Between about 100 and 120°C (corresponding to a depth of about 11,000 feet) all detrital K-feldspar (Or-for orthoclase) is dissolved and all detrital plagioclase is albitized. All components like Ba, Sr, Li, Pb and Zn in these abundant minerals, which make up about 20% of the sand volume (and about 7% of the shale volume[14]), are released to the pore fluids.

oration of seawater results in the precipitation of small amounts of $CaCO_3$ and $CaSO_4$ (actually the mineral gypsum, $CaSO_4 \cdot 2H_2O$) which remove most of the calcium from solution prior to the precipitation of halite (NaCl)[16]. As halite precipitates, sodium and chloride are removed, leaving behind a solution progressively enriched in magnesium, potassium, and sulfate. If such a late-stage seawater-derived solution were to be buried and to react with limestones ($CaCO_3$) to form dolomite ($CaMg(CO_3)_2$), and the sulfate removed by reduction (an unaddressed problem of major proportions because these "bittern" solutions can contain 80 g $SO_4^=$/L), then a Na-Ca-Cl solution would remain (Figure 6). The molar Na/(Ca+Mg+Na) ratio of such a brine is similar to many subsurface formation waters[18].

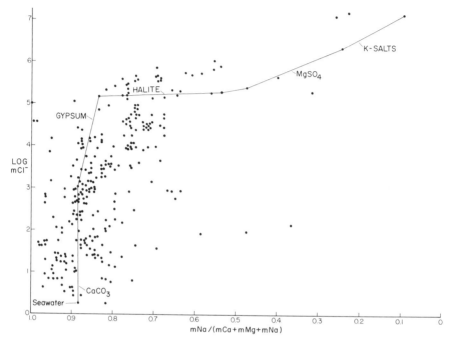

Fig. 6. Molar Na/(Ca+Mg+Na) of saline brines from a variety of basins[17] versus molar chloride concentration. The path for seawater evaporation[16] is also shown, together with the minerals which precipitate at each stage. Almost all natural formation waters can be explained by various mixtures or dilutions of evaporated seawater, but is that the only explanation? Increased calcium concentration with increasing salinity (depth) could merely indicate more rock-water interaction, either with the sediments themselves, or with the underlying basement, in progressively deeper (hotter), more saline (denser) water[17].

Rather than using chloride as a conservative component for comparison of brine compositions, bromide has often been used[16,19]. Bromide is concentrated in seawater-derived brines as halite precipitates, since the distribution coefficient (D) is less than 1 (equation 1).

$$\frac{(Br)}{(Cl)}_{salt} = D \times \frac{(Br)}{(Cl)}_{brine} \qquad (1)$$

Saline brines in sedimentary basins are known to exhibit higher Br/Cl ratios than seawater.

But if halite undergoes recrystallization in contact with a brine, during structural movements for example, equation 1 dictates that recrystallized halite will contain less bromide than the original, and the Br/Cl of the brine will be raised. Measurements of the $^{87}Sr/^{86}Sr$ in anhydrite in Jurassic Louann salt from the Gulf Coast, indicate that most of the anhydrite intimately associated with the impermeable salt has not retained its Jurassic strontium isotopic signature. In fact, radiogenic values evidence extensive reaction with radiogenic basinal pore fluids[20] (Figure 7). If "tight" rocks like massive halite do not preserve chemical signatures derived from primary deposition, then it is difficult to believe that connate brines could preserve such information. It is doubtful if bromide (or any other component) can be treated as a truly conservative component in analyzing brine compositions.

There is no doubt that the major ion composition of most sedimentary basin brines can be derived by mixing and/or diluting various kinds of brines produced by evaporating seawater (Figure 6). But is that how subsurface brines are formed? Can sufficient volumes of surficially generated brine be buried with the sediments, to be later modified by processes such as dolomitization, and dispersed throughout the sedimentary basin? One problem with such a scenario is that evaporites themselves contain essentially no porosity as they are buried. Where in the sedimentary section are such brines "stored" during burial? Adjacent rocks, limestones for example, often contain abundant evidence of extensive alteration by meteoric water soon after deposition, and so they could hardly have acted as passive "containers" for large volumes of connate brine.

The apparent inability to "store" large volumes of

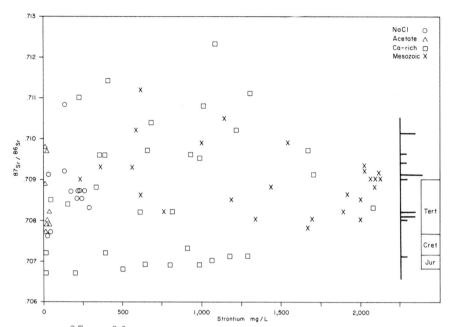

Fig. 7. $^{87}Sr/^{86}Sr$ of brines from the Frio Formation (Oligocene) and Mesozoic formations (x's)[21], Texas Gulf Coast. Seawater during Mesozoic time had an $^{87}Sr/^{86}Sr$ ratio of between 0.7068 and 0.7077, and during the time Frio sediments were being deposited, between about 0.7077 and 0.7082. Most brines not only contain vastly more strontium than could possibly be derived by evaporating seawater, but the strontium is much more radiogenic (contains additional ^{87}Sr) then Jurassic seawater.

connate brine during burial, the need to recycle evaporites, and massive subsurface chemical modification of the rocks themselves (Figure 5), lead me to conclude that subsurface evaporite destruction and subsequent rock-water interaction in the subsurface is more important in controlling the chemistry of subsurface brines than is the original chemistry of surficial brines. Two "classes" of rock-water interaction can be visualized.

The massive destruction of detrital feldspar clearly releases significant amounts of calcium, potassium, strontium, barium, etc. to solution (Figure 5). Feldspar equilibria prove that large amounts of potassium cannot remain in solution until temperatures reach very high values (Figure 4), and potassium is commonly consumed in the formation of diagenetic illitic clay from a smectitic precusor[15]. The strontium and barium contained in saline brines are known to be vastly more abundant than predicted simply by the evaporation of seawater, and the $^{87}Sr/^{86}Sr$ ratio of the brines is commonly elevated (Figure 7). ^{87}Sr is produced by the decay of ^{87}Rb, an element characteristic of silicate phases, especially K-feldspars. Thus significant involvement of silicate phases in determining saline brine chemistry is proven. Other isotope ratios, such as $^{18}O/^{16}O$[22] and D/H[23] are shifted from the values expected for surficial seawater-derived brines, providing additional evidence for extensive rock-water interaction in the subsurface.

In addition to rock-water interaction within the sedimentary basin itself, it is possible that dense brines may interact with the basement beneath the sedimentary basin. It is well known that fluid inclusions in minerals of hydrothermal, metamorphic or igneous origin are commonly very saline[24]. Thus the interaction of saline brines at all temperatures in the Earth's crust is implied.

Metamorphism is accompanied by devolatilization, during which carbon dioxide and water are released, presumably to overlying sediments. The formation of slatey cleavage during low grade metamorphism apparently requires the loss of large volumes of the rocks themselves[25]. The loss of appreciable volumes of insoluble components such as SiO_2 and Al_2O_3 means that they must be transported into overlying stata. In addition to such large scale material transport, the protons which are bound in alumino-silicates during weathering at the Earth's surface are progressively replaced by cations during metamorphism. For example, the H^+/Na^+ of a solution in equilibrium with albite and kaolinite (a clay mineral) increases nearly two orders of magnitude over the temperature interval 25 to $200°C$[12] (Equation 2). In solutions dominated by chloride, weak HCl is thus pro-

$$Al_2Si_2O_5(OH)_4 \text{ (kaolinite)} + 2Na^+ + 4H_4SiO_4 \longrightarrow$$

$$2NaAlSi_3O_8 \text{ (albite)} + 2H^+ + 9H_2O \qquad (2)$$

duced. Acid water lost from metamorphic reactions into the overlying sediments will be neutralized by minerals like calcite, generating Ca-enriched solutions and CO_2. In fact, such a reaction sequence is one possible reason why the CO_2 content of natural gasses increases with increasing depth[26] (Figure 8). The $^{13}C/^{12}C$ in natural gas also tends to become enriched in ^{13}C with increasing CO_2 content as more carbon is apparently derived from inorganic as opposed to organic sources.

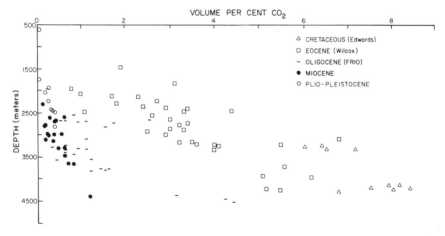

Fig. 8. Volume percent CO_2 in natural gas as a function of depth, Texas Gulf Coast. CO_2 becomes more abundant with increasing depth, and in progressively older formations. A vertical flux of CO_2 from deeper in the basin is one possible explanation. The depth to basement is variable, but is commonly at least 10 kilometers.

Thus rock-water reactions which may be responsible for brine chemistry need not be restricted to the sedimentary basin itself. The interaction of saline brines with basement rocks[27] is as plausible a source for metals like lead and zinc which characterize Mississippi-Valley-Type ore deposits[28] as other sources which have been proposed (redbeds and shales)[29].

CONCLUSION

Density stratification accounts for the regular increase in salinity which is commonly observed in brines in sedimentary basins. The salinity of the water derives primarily from subsurface dissolution of evaporites, principally halite. Halite (or bittern salt) recrystallization accounts for bromide/chloride ratios near or slightly greater than seawater. The calcium content of saline brines is derived from plagioclase albitization, neutralization of HCl generated during metamorphism, and the final expulsion of connate brines which represent the late stage evaporation of seawater. Rock-water interactions which give rise to the chemistry of saline brines are not necessarily restricted to the sedimentary basin itself, but may in part derive by reactions with underlying basement rocks, or the volatile products of metamorphism.

In order to assess the magnitude of subsurface evaporite dissolution and rock-water interaction, future studies of formation water geochemistry must also assess the changes that the rocks themselves have undergone. Only by simultaneous study of the rocks and the waters can the magnitude of rock-water interaction be assessed, and the contribution of connate waters evaluated.

ACKNOWLEDGEMENTS

Numerous individuals and many oil and gas companies could be named whose assistance before, during, and after sampling enabled a large data set to be accumulated and assessed. The assistance of the Geology Foundation of the University of Texas at Austin is also gratefully acknowledged.

REFERENCES

1. P. A. Dickey, "Increasing concentration of subsurface brines with depth," Chem. Geol. $\underline{4}$, 361-370 (1969).
2. D. E. White, "Saline waters of sedimentary rocks," Amer. Assoc. Petrol. Geol. Mem. 4, 342-366 (1965).

3. C. H. Bruce, "Smectite dehydration - its relation to structural development and hydrocarbon accumulation in the Northern Gulf of Mexico basin," Amer. Assoc. Petrol. Geol. Bull. 68, 673-683 (1984); R. A. Morton and L. S. Land, "Regional variations in formation water chemistry, Frio Formation (Oligocene), Texas Gulf Coast," in press, Amer. Assoc. Petrol. Geol. Bull.
4. W. L. Russell, "Subsurface concentration of chloride brines," Amer. Assoc. Petrol. Geol. Bull. 17, 1213-1228 (1933); J. S. Hanor, "Fifty years of development of thought on the origin and evolution of subsurface sedimentary brines," in *Revolution in the Earth sciences: Advances in the past half century*, edited by S. J. Boardman (Kendall/Hunt, Dubuque, Iowa) 99-111 (1983).
5. B. B. Hanshaw, "Cation-exchange constants for clays from electrochemical measurements," Clays and Clay Mins. 12, 397-421 (1964); F. A. F. Berry, "Relative factors influencing membrane filtration," Chem. Geol. 4, 295-301 (1969); D. L. Graf, "Chemical osmosis, reverse chemical osmosis, and the origin of subsurface brines," Geochim. et Cosmochim. Acta. 46, 1431-1448 (1982).
6. F. T. Manheim and M. K. Horn, "Composition of deeper subsurface waters along the Atlantic continental margin," Southeast. Geol. 8, 215-236 (1968); D. C. Bond "Hydrodynamics in deep aquifers of the Illinois basin," Ill. Geol. Surv. Circ. 470, 69 (1972).
7. S. L. Phillips, O. Huseyin, and L. F. Silvester, "Density of sodium chloride solutions at high temperatures and pressures," Lawr. Berk. Labs. 16275, 51 (1983).
8. R. M. Garrels and F. T. Mackenzie, "Evolution of Sedimentary Rocks," (Norton, New York, 1971), p. 272.
9. J. H. Feth, "Chloride in natural continental water - a review," U. S. Geol. Surv. Water Sup. Paper 2176, 30 (1981).
10. L. S. Land and D. R. Prezbindowski, "Chemical constraints and origins of four groups of Gulf Coast reservoir fluids: Discussion," Amer. Assoc. Petrol. Geol. Bull. 69, 119-126 (1985).

11. A. B. Carpenter, "The chemistry of dolomite formation I: The stability of dolomite," in *Concepts and models of Dolomitization*, edited by D. H. Zenger, J. B. Dunham, and R. L. Ethington, Soc. Eco. Paleo. Min. Spec. Pub. 28, 111-122 (1980).
12. H. C. Helgeson, "Chemical interaction of feldspars and aqueous solutions," in *The Feldspars*, edited by W. S. MacKenzie and J. Zussman, Proc. NATO Adv. Study Inst., Manchester U. Press, 184-217 (1972).
13. L. S. Land, "Frio sandstone diagenesis, Texas Gulf Coast: A regional isotopic study," in *Clastic Diagenesis*, edited by D. A. McDonald and R. C. Surdam, Amer. Assoc. Petrol. Geol. Mem. 37, 47-62 (1984).
14. J. E. Hower, E. Eslinger, M. E. Hower, and E. A. Perry, "Mechanism of burial metamorphism of argillaceous sediments: I. Mineralogical and chemical evidence," Geol. Soc. Amer. Bull. $\underline{87}$, 725-737 (1976).
15. V. Schmidt and D. A. McDonald, "The role of secondary porosity in the course of sandstone diagenesis," in *Aspects of Diagenesis*, edited by P. A. Scholle and R. Schluger, Soc. Eco. Paleo. Min. Spec. Pub. 26, 175-208 (1979).
16. A. B. Carpenter, "Origin and chemical evolution of brines in sedimentary basins," in *13th Ann. Forum on the geology of industrial minerals*, edited by K. S. Johnson and J. Russell, Okla. Geol. Surv. Circ. 79, 60-77 (1978).
17. P. Fritz and S. K. Frape, "Saline groundwaters in the Canadian sheld - A first overview," Chem. Geol. $\underline{36}$, 179-190 (1982).
18. D. L. Graf, W. F. Meents, I. Friedman, and N. F. Shimp, "The origin of saline formation waters III: Calcium chloride waters," Ill. State Geol. Surv. Circ. 397, 60 (1966); A. B. Carpenter, M. L. Trout, and E. E. Picket, "Preliminary report on the origin and chemical evolution of lead- and zinc-rich oil field brines in central Mississippi," Econ. Geol. $\underline{69}$, 1191-1206 (1974); L. S. Land and D. R. Prezbindowski, "The origin and evolution of saline formation water, Lower Cretaceous carbonates, south-central Texas," Jour. Hydrology $\underline{54}$, 51-74 (1981); Y.

K. Kharaka, E. Callender, and W. W. Carothers, "Geochemistry of geopressured geothermal waters from the Texas Gulf Coast," Proc. 3rd Geopr.- Geoth. Energy Conf., Lafayette La. 2, GI121-GI164 (1977); Problems in petroleum geology, edited by W. E. Wrather and F. H. Lahee, Amer. Assoc. Petrol. Geol., pt. VI (1934).

19. G. Rittenhouse, "Bromine in oil field waters and its use in determining possibilities of origin of these waters," Amer. Assoc. Petrol. Geol. Bull. 51, 2430-2440 (1967).

20. H. H. Posey, "Regional characteristics of strontium carbon and oxygen isotopes in the salt dome cap rocks of the western Gulf Coast," Ph.D. Thesis, N. Carolina - Chapel Hill, 248 (1986).

21. A. M. Stueber, P. Pushkar, and E. A. Heatherington, "A strontium isotopic study of Smackover brines and associated solids, southern Arkansas," Geochim. et Cosmochim. Acta 48, 1637-1649 (1984).

22. R. N. Clayton, I. Friedman, D. L. Graf, T. K. Mayeda, W. F. Meents, and N. F. Shimp, "The origin of saline formation waters, I. Isotopic composition," Jour. Geophys. Res. 71, 3869-3882 (1966).

23. H.-W. Yeh, "D/H ratios and late-stage dehydration of shales during burial," Geochim. et Cosmochim. Acta. 44, 341-352 (1980).

24. E. Roedder, "Fluid inclusions," Min. Soc. America Rev. in Min. 12, 644 (1984).

25. E. C. Buetner and E. G. Charles, "Large volume loss during cleavage formation, Hamburg sequence, Pennsylvania," Geol. 13, 803-805 (1985); T. O. Wright and L. B. Platt, "Pressure solution and cleavage in the Martinsburg shale," Amer. Jour. Sci. 282, 122-135 (1982).

26. P. D. Lundegard and L. S. Land, "Carbon dioxide and organic acids: Their role in porosity enhancement and cementation, Paleogene of the Texas Gulf Coast," in Roles of organic matter in sediment diagenesis, edited by D. L. Gautier, Soc. Econ. Paleo. Min. Spec. Pub. 38, 129-146 (1986).

27. Y. N. Shieh, "Oxygen isotope study of pre-cambrian granites from the Illinois deep hole project," Jour. Geophys. Res. 88, 7300-7304 (1983).

28. B. R. Doe, J. S. Stuckless, and M. H. Delevaux, "The possible bearing of the granite of the UPH deep drill holes, northern Illinois, on the origin of Mississippi valley ore deposits," Jour. Geophys. Res. 88, 7335-7345 (1983).
29. A. B. Carpenter, "Interim report on lead and zinc in oil-field brines in the central Gulf coast and in southern Michigan," Soc. Mining Engin. of AIME Preprint 79-95,15 (1979).

ENERGETICS OF COMPLEX ALUMINOSILICATES

Alexandra Navrotsky
Dept. of Geological and Geophysical Sciences
Princeton University, Princeton NJ 08544

ABSTRACT

Systematic trends seen in the thermodynamics of aluminosilicate crystals and glasses at high temperature may be applicable to the complex aluminosilicates encountered as porous media. Specific features of the energetics of charge-coupled substitutions (e.g. $Si^{4+} \rightarrow Al^{3+}+Na^{+}$), of Al,Si order-disorder, and of the many metastable states shown by crystalline and amorphous SiO_2 are discussed.

INTRODUCTION

To a solid state chemist accustomed to dealing with well crystallized solids and homogeneous glasses at high temperature, porous media present a challenging complexity of crystallographically ill-defined, heterogeneous, and non-equilibrium assemblages. Yet their backbone is often an aluminosilicate sheet or framework, and local atomic environments in these materials may be quite similar to those in simpler high temperature phases. The goal of this paper is to summarize some thermodynamic and structural trends seen in high temperature aluminosilicates and to ask whether such systematics offer any insight into the chemical behavior of clays, zeolites and other low temperature materials. This paper will pose many questions and few answers.

CHARGE-COUPLED SUBSTITUTIONS

Starting with SiO_2 (either crystalline or amorphous) one can maintain a nominally completely polymerized framework and substitute tetrahedral cations of lower charge (T^{3+} = Al,Ga,Fe,B) with charge compensation outside the framework by other cations (A^+ = alkali, A^{2+} = alkaline earth or transition metal). The reaction may be written as

$$(1-x)\ SiO_2 + xA^{n+}_{1/n}T^{3+}O_2 \rightarrow A^{n+}_{x/n}(T^{3+}_x Si^{4+}_{1-x})O_2 \qquad (1)$$

In the glassy state, this substitution can occur continuously, though it is often limited at $x > 0.5$ by the inability to quench a glassy material from a melt. In crystals, specific compounds with different framework topologies (β-quartz, feldspar, nepheline, cordierite structures) are formed, which usually have fairly small solid solution ranges.

From an energetic viewpoint, several questions can be formulated. For a given structure, how does the energy of a given type of substitution depend on the nature of the substituting cation? Can this variation be explained in terms of crystal chemical and bonding factors? For different structures, do analogous substitutions have similar energetics? Can such trends be used to predict thermodynamic properties, especially of complex sheet and chain silicates? This section will summarize some preliminary answers to these questions. Much experimental and theoretical work still needs to be done.

A series of aluminosilicate glasses $SiO_2-A^{n+}_{1/n}AlO_2$ (A = alkali or alkaline earth) have been studied by high temperature solution calorimetry (1,2,3). For aluminosilicate glasses, both x-ray scattering and Raman spectroscopy suggest that, when Al substitutes in a charge-coupled fashion for Si, an aluminosilicate framework is maintained although there may be changes in regularity (increased bond angle and ring size variation) with Al content. The extent of Al,Si ordering (Al-avoidance) in these glasses is still controversial (1,4,5).

The perturbation of the framework resulting from the substitution also depends strongly on the nature of the charge balancing cation, greater disturbance occurring with ions of smaller size and large charge. The heat of solution data are shown in Figure 1. The process of dissolving an aluminosilicate glass in

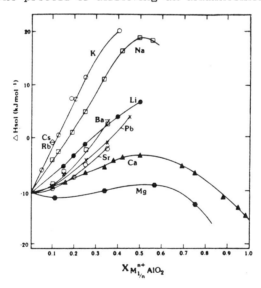

Figure 1. Enthalpies of solution of glasses $Al^{n+}_{1/n}AlO_2-SiO_2$ in molten $2PbO \cdot B_2O_3$ at 973 K.

molten lead borate to form a dilute (<1 wt %) solution consists of breaking up an aluminosilicate framework into isolated species, presumably SiO_4 and AlO_4 tetrahedra and alkali and alkaline-earth cationic species dissolved in a borate matrix. Thus the enthalpy of solution may be considered a measure of the strength of bonding in the aluminosilicate glass, at least in a relative sense when comparing various compositions. Three points are evident. First, the enthalpies of solution generally become more endothermic as $A^{n+}_{1/n}AlO_2$ is substituted for SiO_2, for x < 0.5. This increase becomes more pronounced with decreasing field strength (or increasing basicity) of the cation, e.g., in the series Mg, Ca, Sr, Pb, Ba, Li, Na, K, Rb, Cs. Second, the enthalpy of solution relations tend to curve back toward more exothermic values, with a maximum near x = 0.5. This maximum reflects an exothermic enthalpy of mixing for reaction (1). This enthalpy of mixing becomes more exothermic with increasing basicity of the metal oxide or decreasing ability of the M cation to bond to oxygen. Third, at 0 < x < 0.4, pronounced curvature occurs in the relations for the alkaline-earths while the relations for the alkalis are approximately linear. This curvature implies a less exothermic heat of mixing in this region and may presage glass-glass metastable immiscibility analogous to but less pronounced than the immiscibility seen in binary metal oxide-silica systems.

The calorimetric data confirm regular systematics for the entire alkali and alkaline earth series. The enthalpy of the substitution may be measured by

$$\Delta H(stab) = [-\Delta H_{sol}(SiO_2) + \Delta H_{sol}(A_{x/n}Al_xSi_{1-x}O_2)]/x. \quad (2)$$

Values of $\Delta H(stab)$ are plotted in Figure 2 against the ionic

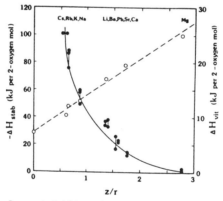

Figure 2. Enthalpy stabilization of glasses (Eqn. 3) plotted against ionic potential (z/r) of cation A (left-hand scale and curve). Enthalpy of vitrification (right-hand scale and line).

potential, z/r, of the cation, with z the formal charge and r the Shannon and Prewitt ionic radius using coordination number 6 for Li and Mg, 8 for the other ions. Though the choice of charge and size is rather arbitrary, the ionic potential offers a reasonable parameter, easily calculated for all ions, for comparing the relative strength of bonding of different cations. The stabilization of the charge coupled substitution in aluminosilicate glasses is inversely related to the ability of A to bond to oxygen.

Ab initio molecular orbital calculations on aluminosilicate clusters chosen to model the above systems have recently been completed (6,7). This approach chooses small molecular clusters to mimic the local environments of atoms within a solid. The methods of computational quantum chemistry are used to solve the Schrodinger Equation and find eigenvalues for physical parameters such as bond lengths, bond angles, total energies and charge distributions. Different levels of approximation, ranging from empirical to semi-empirical to almost completely ab initio, are used to describe the wave functions. Fortunately, for framework silicates, the light elements Si, Al, and O are major constituents, and the wave functions of molecules containing these atoms and hydrogens to terminate bonds in the cluster can be calculated fairly rigorously. Thus in the problem above, the basic cluster, which contains an SiOAl linkage (bridging oxygen), is an H_6AlSiO_7 molecule. The bridging oxygen is then approached by a metal atom A (Li, Na, Mg, Al) at a bond length characteristic of the AO bond in minerals. This results in clusters of the type $A(OH)_3-H_6T_2O_7$ and $A(H_2O)_5-H_6T_2O_7$ for which STO-3G molecular orbital calculations result in the following conclusions. When an octahedrally or tetrahedrally coordinated cation cluster (A = Li, Na, Mg) is attached to the bridging oxygen of an SiOSi or AlOSi linkage, the TO bonds are perturbed such that the TOT angle narrows and the bridging TO bonds lengthen slightly. This perturbation increases in the order Na, Li, Mg. The bridging AlO bond is lengthened more than the bridging SiO bond. The effects of tetrahedrally and octahedrally coordinated A atoms (at their "normal" AO distances) are generally comparable. Calculated and experimental SiO and AlO bond lengths are in good agreement for ordered framework aluminosilicate crystals containing Li, Mg, and Na.

The thermochemical data for glasses can be correlated with the changes (both calculated and observed) in TO bond lengths mentioned above (7). The average perturbation of the bridging TO bond length is calculated as follows. The AlO and SiO distances in the ordered crystals for those oxygens bonded to two tetrahedrally bonded framework cations plus at least one other nonframework cation can be compared to the "ideal" TO bond lengths of 1.712 (AlO) and 1.581 (SiO) taken as the average of SiO and AlO bond lengths in framework aluminosilicates for which the oxygen is only 2-coordinate. Thus

$$\Delta(AlO) = (AlO)_{obs} - 1.712, \qquad (3)$$

$$\Delta(SiO) = (SiO)_{obs} - 1.581, \qquad (4)$$

and, at a composition with $Al/(Al+Si) = x$ and $Si/(Al+Si) = 1-x$,

$$\Delta(TO)_{av} = x\ \Delta(AlO) + (1-x)\ \Delta(SiO) \qquad (5)$$

The resulting plot (Fig. 3) shows strong correlation between $\Delta H(stab)$ and $\Delta(TO)$. Thus one can parameterize the perturbation of the aluminosilicate framework using a bond length perturbation parameter consistent with molecular orbital calculations on a covalently bonded cluster or by using an essentially ionic field strength parameter, z/r.

The enthalpy difference between crystal and glass (enthalpy of vitrification, $\Delta H(vit)$, shows a linear correlation with z/r or $\Delta(TO)$

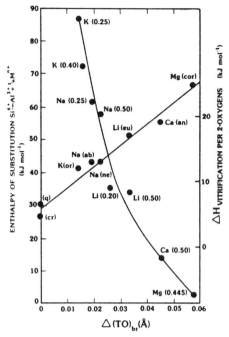

Figure 3. Enthalpy stabilization (left-hand scale and curve) and enthalpy of vitrification (right-hand scale and line) plotted against average tetrahedral bond length perturbation parameter $\Delta(TO)$

(see Figs. 2 and 3). If one considers silica (quartz and glass) as a framework structure with the interstices occupied by a cation (i.e., a vacancy) of infinitely weak ability to perturb the bridging oxygen, then the point for SiO_2 with $z/r = 0$ and $\Delta(TO) = 0$ lies on the same trend as the other data. The heat of vitrification increases with increasing z/r or $\Delta(TO)$, indicating that increasing perturbation of the aluminosilicate framework not only decreases the stability of the glass with respect to mixing properties in the amorphous state but also with respect to the crystalline state. Similar trends probably hold for the molten state as for the glass, but discussion of the relations among glass and melt properties is beyond the scope of this chapter.

To compare the same substitution in different crystalline and glassy systems, Table 1 lists the energetics of a number of substitutions, calculated from heat of solution data in a manner analogous to that for the enthalpy stabilization of aluminosilicate glasses discussed above. The table reveals several regularities. For the substitution, $Si^{4+}_T \rightarrow Al^{3+}_T + 1/nA^{n+}_A$, the enthalpy becomes less exothermic with increasing z/r of the cation for crystals as well as for glasses, with values of (negative) 110-120 kJ for Cs and Rb, 80-100 kJ for K, 60-85 kJ for Na, near 40 kJ for Li, and substantially less exothermic values for the alkaline earths. For the alkalis, values generally overlap for glasses and crystals and span a range of about ±10% around of the average value. Thus, to within an accuracy of about ±10%, it appears that the enthalpy of the coupled silica-stuffing substitution is constant in different long range environments (e.g., for Na; glass, feldspar, nepheline,

Table 1

Enthalpies of Substitution Reactions in Silicates [8]

Substitution and Mineral Pair[a]	ΔH (kJ mol^{-1})
$Si^{4+}_T \rightarrow Al^{3+}_T + Cs^+_A$	
SiO_2(glass)-$Cs_{0.1}Al_{0.1}Si_{0.9}O_2$(glass)	-112
$Si^{4+}_T \rightarrow Al^{3+}_T + Rb^+_A$	
SiO_2(glass)-$Rb_{0.1}Al_{0.1}Si_{0.9}O_2$(glass)	-115
SiO_2(quartz)-$Rb_{0.25}Al_{0.25}Si_{0.75}O_2$(microcline)	-128
$Si^{4+}_T \rightarrow Al^{3+}_T + K^+_A$	
SiO_2(glass)-$K_{0.25}Al_{0.25}Si_{0.75}O_2$(glass)	-88
SiO_2(quartz)-$K_{0.25}Al_{0.25}Si_{0.75}O_2$(high sanidine)	-100
SiO_2(quartz)-$K_{0.33}Al_{0.33}S_{0.67}O_2$(leucite)	-82
SiO_2(quartz)-$K_{0.5}Al_{0.5}Si_{0.5}O_2$(high kalsilite)	-78
$Si^{4+}_T \rightarrow Al^{3+}_T + Na^+_A$	
SiO_2(glass)-$Na_{0.25}Al_{0.25}Si_{0.75}O_2$(glass)	-61
SiO_2(quartz)-$Na_{0.25}Al_{0.25}Si_{0.75}O_2$(high albite)	-85
SiO_2(quartz)-$Na_{0.5}Al_{0.5}Si_{0.5}O_2$(nepheline)	-70
$Ca_2Mg_5Si_8O_{22}F_2$(F-tremolite)-$NaCa_2Mg_5AlSi_7O_{22}F_2$ (F-edenite)	-79
$Si^{4+}_T \rightarrow Al^{3+}_T + Li^+_A$ (or M)	
SiO_2(glass)-$Li_{0.25}Al_{0.25}Si_{0.75}O_2$(glass)	-37
SiO_2(quartz)-$Li_{0.5}Al_{0.5}Si_{0.5}O_2$($\beta$-eucryptite)	-41
$Ge^{4+}_T \rightarrow Al^{3+}_T + Na^+_A$	
GeO_2(glass)-$Na_{0.25}Al_{0.25}Ge_{0.75}O_2$(albite)	-105
GeO_2(quartz)-$Na_{0.25}Al_{0.25}Ge_{0.75}O_2$(albite)	-85
$Si^{4+}_T \rightarrow Al^{3+}_T + 1/2Mg^{2+}_A$	
SiO_2(glass)-$Mg_{0.222}Al_{0.444}Si_{0.556}O_2$(glass)	~ 0
SiO_2(quartz)-$Mg_{0.222}Al_{0.444}Si_{0.556}O_2$($\beta$-quartz)	+ 3.1
$Si^{4+}_T \rightarrow Al^{3+}_T + 1/2Mg^{2+}_M$	
SiO_2(quartz)-$Mg_{0.222}Al_{0.444}Si_{0.556}O_2$(cordierite)	-25 to -36
$Si^{4+}_T \rightarrow Al^{3+}_T + 1/2Ca^{2+}_A$	
SiO_2(glass)-$Ca_{0.25}Al_{0.5}Si_{0.5}O_2$(glass)	-14
SiO_2(quartz)-$Ca_{0.25}Al_{0.5}Si_{0.5}O_2$(anorthite)	-38
$Si^{4+}_T \rightarrow Al^{3+}_T + 1/2Sr^{2+}_A$	
SiO_2(glass)-$Sr_{0.125}Al_{0.25}Si_{0.75}O_2$(glass)	-22
$Si^{4+}_T \rightarrow Al^{3+}_T + 1/2Ba^{2+}_A$	
SiO_2(glass)-$Ba_{0.125}Al_{0.25}Si_{0.75}O_2$(glass)	-33[b]

amphibole). This suggests that such substitutions are controlled energetically primarily by the local environment, and that as long as the framework (T) and interstitial (A) sites remain reasonably similar, the energy of the substitution remains rather constant. For the alkaline earths (Mg and Ca), the difference between crystal and glass becomes more pronounced (although Al,Si order in anorthite and cordierite may be a complicating feature). The substitution which puts Mg into octahedral sites in cordierite is energetically much more favorable than that which places it into large interstitial sites in the stuffed β-quartz structure or in glass.

Can the above generalizations be extended to other complex aluminosilicates? A starting point might be to test whether the reaction $Si_T^{4+} \rightarrow A_A^+ + Al_A^{3+}$ (A = Na, K) has similar energetics in micas, clays and zeolites as it appears to have in feldspars, glass, and amphiboles. If so, then the thermodynamic properties of a variety of low temperature phases could be estimated. A second approach would be to try to correlate framework topology (TOT angles and ring and cavity sizes) with the ionic potential of the charge compensating mono- or divalent cation in a quantitative fashion. The stability of such a framework on dehydration may also relate to such interactions. For such thermochemical and structural studies, materials which are extremely homogeneous and well-characterized are essential.

ALUMINUM-SILICON ORDER-DISORDER

The energetics of Al, Si order-disorder in cordierite (9) and the plagioclase feldspars (10) have been studied recently and compared with structural and spectroscopic studies to correlate structural and energetic changes. Glass of cordierite, $Mg_2Al_4Si_5O_{18}$, composition crystallizes rapidly at 1200 or 1400 °C to a highly Al,Si disordered cordierite. On annealing (see Fig. 4) this

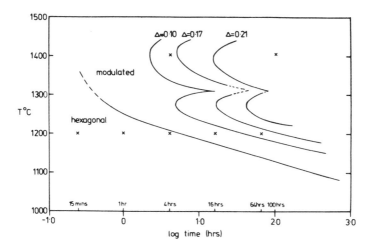

Figure. 4. Time, temperature, transformation plot for cordierite crystallizing from glass (9).

hexagonal phase develops structural modulations discernable by
electron microscopy and finally develops macroscopic deviations
from hexagonal symmetry characterized by a distortion index, Δ.
The enthalpy of solution of this material becomes more endothermic

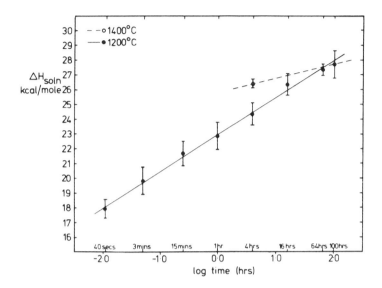

Figure 5. Enthalpy of solution in lead borate at 700 °C of
cordierite crystallized from glass and annealed at 1200 or 1400 °C
for various times (9).

in a linear fashion with log(time) (see Fig. 5) indicating that the
energetic stabilization arising from Si-Al ordering increases in a
very regular fashion with annealing and does not show any
detectable discrete changes as the structure progresses from
hexagonal to modulated to twinned orthorhombic. This behavior
suggests that the energy of the Al, Si interchange step is
independent of the degree of local or long range order. This
assumption leads to a relationship betwen the rate of change of
wrongly occupied sites (dN/dt) and the number remaining wrongly
occupied (N). The rate of ordering drops off as N decreases, but
in a much more rapid manner than for radioactive decay. These
empirical relations are not fully understood, though there may be
some affinity with low temperature creep in metallic systems for
which an analogous expression (strain ∝ log time) is commonly
found to hold and in which the net effect (i.e. overall strain) is
made up of a very large number of small increments. There are
two immediate implications, however. Firstly, no distinction can be
made between the energetics effects of long range ordering and
short range ordering. By the time the cordierite is detectably
orthorhombic it is substantially ordered. Secondly, for the order
modulations to grow and coarsen into twins, some Al/Si pairs must
re-exchange, since they might be correctly distributed with respect
to their local domain but incorrectly distributed relative to the
domain of which they become a part during the coarsening. Thus

the actual atomic picture is of a large number of Al-Si exchanges occurring such that a continuous rearrangement of atoms is achieved, with a net drift towards improved ordering.

The use of enthalpy as a measure of reaction progress for an order/disorder transformation is not usual. The only other similar kinetic data known to the authors relate to the disordering of Al and Si in albite feldspar. Holm and Kleppa measured heats of solution of albite crystals which had been disordered at 1045 ± 5 °C for times of up to 35 days. Their data are reproduced in Fig. 6a, showing an apparent step at ~15-20 days which they interpreted as implying a two stage mechanism of disordering. In Fig. 6b, the enthalpy of solution is replotted against log annealing time;

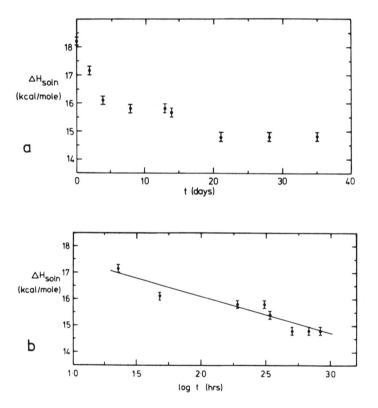

Figure 6. (a) Enthalpy of solution data for albite annealed at 1045 ± 5 °C; (b) The same data plotted against the logarithm of annealing time.

the data are almost consistent with a straight line fit similar to Fig. 5. The same kind of continuous kinetic behavior, as has been outlined in cordierite under metastable conditions, may also apply, therefore, to disordering in albite. Note that the albite study is of a disordering reaction while the cordierite study is of ordering. Thus similar rate laws may apply to both.

Carpenter et al. (10) studied the energetics of order-disorder in the plagioclase feldspars ($NaAlSi_3O_8$-$CaAl_2Si_2O_8$). Estimated enthalpy changes for several structural changes are summarized in Table 2.

Table 2.

Reaction	$\Delta H°$ (kcal/mole)
low albite → high albite	3
"e" plagioclase → $C\bar{1}$ high albite structure	1.4 to 2.8
$I\bar{1}$ → $C\bar{1}$ high albite structure	0.7 to 1.9
$I\bar{1}$ equilibrated at low T → $I\bar{1}$ equilibrated at high T	0.8 to 1.8

The enthalpy of mixing data are consistent with an interpretation of the solid solution as being composed, at high temperatures, of two ideal (zero heat of mixing) segments, one with $C\bar{1}$ symmetry and one with $I\bar{1}$ symmetry, and having a non-first order (continuous) order/disorder transformation between them. The low temperature series can also be separated into two distinct trends, for $I\bar{1}$ and "e" structures.

Values of the enthalpy change due to disordering show a number of systematic trends. Firstly, the values for "e" → $C\bar{1}$ are larger than for $I\bar{1}$ → $C\bar{1}$ in the composition range where both "e" and $I\bar{1}$ structures are observed (~An_{65}-An_{72}). Secondly, the enthalpy change on disordering the most ordered "e" structures at An-rich compositions is larger than for Ab-rich "e" structures. Thirdly, the large enthalpy change of the "e" structure, due to ordering, may be sufficient to stabilize it relative even to a mixture of low albite plus anorthite. It is apparent that all the structures can be quite competitive in terms of free energy and that local and long-range order interact in complex ways.

In terms of applications to low temperature materials, several questions are suggested by the above trends. Is the Al,Si interchange step similar, leading to analogous order-disorder kinetics and ΔH versus log T behavior? Do apparently disordered materials contain substantial short range order? Are various modulated and incommensurate structures intermediates in the order-disorder process? Do such structures and the order-disorder process itself affect physical, surface, and catalytic properties? A combination of careful synthesis, crystallography, calorimetry, and vibrational and NMR spectroscopy would be very fruitful for such materials.

METASTABLE FORMS OF SiO_2

Combined calorimetric and structural studies on chemically vapor deposited SiO_2-based films have shown the following. CVD SiO_2 is energetically metastable relative to fused SiO_2 glass by up to 10 kcal/mol (12,13). The extent of metastability increases with decreasing deposition temperature and decreases with thermal annealing (see Fig. 7). The excess enthalpy appears to consist of two contributions, one which anneals rapidly at 700°C and another which requires complete bond breaking (as during dissolution in the calorimetric solvent) to be released. An energetic state approaching that of bulk silica glass is approached on annealing at

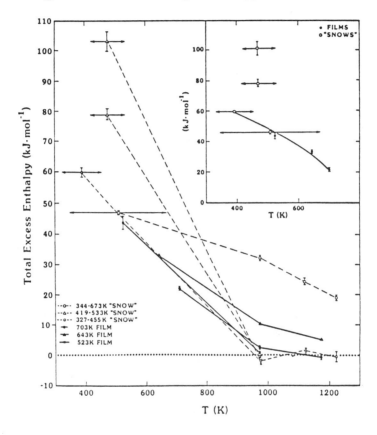

Figure 7. Total excess enthalpies, relative to fused silica, of CVD silica films and "snows" plotted aginst annealing temperature (24 hours) for samples deposited at indicated temperatures (13).

800-900°C, though the crystallization of cristobalite intervenes at the higher temperatures. Infrared and Raman spectra show the initial presence and gradual annealing of SiH and SiOH species in the CVD films and powders (13,14). The vibrational spectra also suggest subtle changes in the tetrahedral network. X-ray scattering studies of the CVD amorphous SiO_2 powder ("snow") (15) show it to contain a distribution of voids which, however, do not bring enough atoms near the internal surfaces to account for the energetics. The radial distribution function (RDF) of the CVD powder and ordinary silica glass show significant differences (see Fig. 8). The RDF of "snow" powder shows much less structure beyond 1 nm, suggesting greater disorder in ring size distribution. The Si-O, O-O, and Si-Si distances show a significantly greater variance (after correction for thermal vibration), suggesting greater disorder in bond lengths and angles. A simple calculation of the extra energy associated with

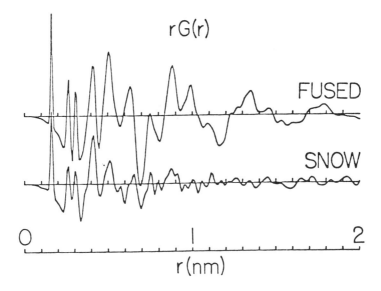

Figure 8. Radial distribution function (RDF) of "snow" and bulk SiO_2

this greater variation, based on conventional force constants, gives a value within 10% of the calorimetric observation.

The existence of a large number of crystalline phases and of distinct amorphous materials all of composition SiO_2 offers an opportunity to formulate a number of fundamental questions.

The energy of a phase is determined in part by its density. In SiO_2 the 3-dimensional connected tetrahedral framework is limited at high densities by a phase (stishovite) with octahedrally coordinated silicon. The highest density tetrahedral phase is coesite, which contains SiOSi angles of 180°. Glasses of high density can be prepared under high pressure. How do their energies and topologies relate to that of coesite? At lower than normal densities, amorphous films, "snows", sol-gel products, flame-oxidation products, zeolites and clathrasils can be prepared (as well as the high temperature polymorphs tridymite and cristobalite, and fusion-formed-glass). Is there a limit of low density at which network connectivity is lost? Is there an analogous limit in energetics?

The available data are summarized in Fig. 9 which shows the enthalpy (relative to quartz) versus molar volume for crystalline and amorphous SiO_2 phases. Similar curves are seen, with the minimum energy structure occurring at a lower density for amorphous phases. The point for stishovite (the octahedrally coordinated dense crystalline phase) falls on a reasonable extrapolation of the data for tetrahedral phases. The enthalpy of silicalite (16), a zeolitic silica with low density and large internal

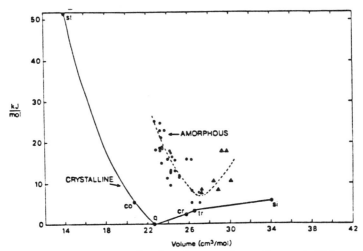

Figure 9. Enthalpy, relative to quartz, of crystalline and amorphous forms of SiO_2 versus their molar volume.

pores (17), is perhaps less destabilizing than one might expect from a smoothly curved extrapolation of the data for quartz, cristobalite, and tridymite. Perhaps once rather large pores are present, one must distinguish between the average density of the framework and of the pores. More work on other low-density crystalline clathrasil (18) and zeolitic (18,19) silicas would help answer such questions.

The data for amorphous phases are from three sources. The cross is "normal" bulk fusion-formed glass, the open circles are pressure densified glasses (20) (with considerable scatter in the energetic data), the triangles are taken as the "structural" portion of the excess enthalpy of CVD films, that portion of the energy that is released only when the materials are dissolved in lead borate at 700 °C (13). The densities of those films are poorly known, uncertainties of ±1 cm³/mole exist in the molar volumes. No density data could be attained for "snow" CVD silicas so those thermochemical data could not be used. Despite these uncertainties, one sees a roughly symmetrical curve of energy versus deviation from normal volume. Further work on a broader range of better characterized samples is needed. Structural models for the amorphous materials of varying density need to be developed.

From a molecular-orbital point of view, a good deal of effort has gone into characterizing a potential energy surface, energy vs. SiO bond distance and versus SiOSi angle (the latter reflecting ring size distribution in a tetrahedral framework) (21). Do the energetics of phases and microstates follow systematic trends suggested by such calculations and by bond stretching and angle bending force constants consistent with this view of bonding? The calculations of the excess energy of "snow" SiO_2, based on the SiO, OO, and SiSi distributions seen in the RDF suggest that such an approach is useful for comparing various amorphous silicas.

A correlation between SiO bond lengths and SiOSi angles has

been suggested by Gibbs and coworkers both on the basis of empirical observation and ab initio molecular orbital calculations (21). A "potential energy surface" which contours energy versus SiO distance and SiOSi angle (21) is shown in Fig. 10. Such a

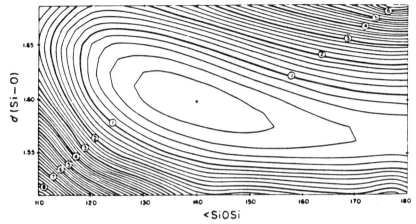

Figure 10. Potential energy surface for $H_6Si_2O_7$ molecule, showing equal-energy contours relative to minimum shown as cross. Dark contours represent interval of 13 kJ/mole (21).

relation gives clues to energetics, kinetics, and vibrations. Can one further quantify the observed patterns and find the energetic cost of the wider distribuitons of local and mid-range geometries in low-density and high-density silicas? Do similar arguments hold for transition states and kinetic features?

To what extent can the effects of a large surface area be separated from those of structural changes in a "bulk" material? Can surface effects and those of H_2O content be separated? How does annealing affect surface and bulk structure and energetics? Systematic energetic studies combined with structural characterization along time-temperature-"transformation" traverses would be very useful.

ACKNOWLEDGEMENTS

The work summarized here has been supported by the National Science Foundation (DMR 821127, EAR 8405218, 8513916) and the Department of Energy (DEFCO285ER13437).

REFERENCES

(1) A. Navrotsky, G. Peraudeau, P. McMillan and J. P. Coutures, Geochim. Cosmochim. Acta 46, 2039 (1982).
(2) R. L. Hervig and A. Navrotsky, Geochim. Cosmochim. Acta 48, 513 (1984).
(3) B. N. Roy and A. Navrotsky, J. Am. Ceram. Soc. 67, 606 (1984).
(4) D. J. Henry, A. Navrotsky, and H. D. Zimmermann, Geochim. Cosmochim. Acta 46, 381 (1982).

(5) J. B. Murdoch, J. F. Stebbins, and I. S. E. Carmichael, Am. Min. 70, 332 (1985).
(6) K. L. Geisinger, G. V. Gibbs, and A. Navrotsky, Phys. Chem. Min. 11, 266 (1985).
(7) A. Navrotsky, K. L. Geisinger, P. McMillan, and G. V. Gibbs, Phys. Chem. Min. 11, 284 (1985).
(8) A. Navrotsky in "Microscopic to Macroscopic, Reviews in Mineralogy, Vol. 14", S. W. Kieffer and A. Navrotsky, Eds., (Min. Soc. Amer., Washington, DC, 1985), p. 225.
(9) M. A. Carpenter, A. Navrotsky, and J. D. C. McConnell, Geochim. Cosmochim. Acta 47, 899 (1983).
(10) M. A. Carpenter, A. Navrotsky and J. D. C. McConnell, Geochim. Cosmochim. Acta 49, 947 (1985).
(11) J. L. Holm and O. J. Kleppa, Am. Min. 53, 123 (1968).
(12) M. Huffman, Ph.D. Thesis, Arizona State University, Aug. 1985.
(13) M. Huffman, A. Navrotsky, and F. S. Pintchovski, J. Electrochem. Soc. 133, 164 (1986).
(14) M. Huffman and P. McMillan, Jour. Noncryst. Solids 76, 369 (1985).
(15) J. Konnert, P. D'Antonio, M. Huffman, and A. Navrotsky, Jour. Amer. Ceram. Soc. (in press).
(16) G. K. Johnson, I. R. Tasker, D. A. Howell, and J. V. Smith, Jour. Chem. Thermo. (in press).
(17) E. M. Flanigan, J. M. Bennett, R. W. Grose, J. P. Cohen, R. L. Patton, R. M. Kirchner, and J. V. Smith, Nature 271, 512 (1978).
(18) F. Liebau, Zeolites, 3, 191 (1983).
(19) J. M. Thomas and J. Klinowski, Advances in Catalysis 33, 199 (1985).
(20) R. Couty, Ph.D. Thesis, Univ. Paris VI (1977).
(21) G. V. Gibbs, Am. Min. 67, 421 (1982).

EXPERIMENTAL STUDIES ON PHYSICAL PROPERTIES OF VERMICULITE, MUSCOVITE AND KAOLINITE CLAYS

N. Wada
Schlumberger-Doll Research, Ridgefield, CT 06877-4108

ABSTRACT

In this paper, we will focus on intercalation of molecules (especially water) and accompanying structural changes in well-characterized clays. Experimental results obtained by such methods as isotherm water adsorption, magnetic susceptibility, x-ray scattering, neutron scattering, Raman scattering, high pressure etc., will be presented to discuss hydration states, magnetic properties, lattice dynamics and pressure effects on clays.

INTRODUCTION

Although clays (layered silicates) are commonly present in porous rocks and play a crucial role in determining the physical and chemical properties of rocks, the fundamental properties of clays are not yet well understood. Isomorphic substitutions and crystalline disorder usually make it difficult to obtain a comprehensive understanding of clays. In this paper, we present experimental work on *well-characterized* clays. These fundamental results from specific but well-characterized clays should contribute to a better understanding of ordinary clays in rocks, and thus allow us to have a deeper insight into porous rocks.

STRUCTURE

Clays are made of two basic sheets; one is a Si-tetrahedral sheet, and the other is an Al- or Mg-octahedral sheet.[1] Depending on the stacking sequence, they are classified into three classes; 2:1 clays consist of two tetrahedral sheets with an octahedral sheet in-between, 1:1 clays consist of one tetrahedral and one octahedral sheets, and 2:1:1 clays accommodate an octahedral sheet between the 2:1 layers.

Vermiculite is an expandable 2:1 clay, which exhibits well-defined hydration states. In vermiculite, some of the Si^{4+} atoms in the tetrahedral sites are replaced with Al^{3+} atoms. This causes a net charge deficiency in the silicate layer. In order to compensate this charge deficiency, cations are accommodated between the silicate layers. Those cations can be exchanged easily by immersing vermiculite in a salt solution. We prepared Na-, Ni- and Sr-vermiculite samples from single-crystal Mg-vermiculite from Llano county in Texas.

Muscovite is also a 2:1 clay, but is not expandable like vermiculite. The interlamellar cations are potassium atoms which balance the charge deficiency due to the substituted Al^{3+} atoms in the Si^{4+} sites. The net interlamellar charge per unit area is more in muscovite than vermiculite. Muscovite contains Al-dioctahedral layers, while Llano vermiculite essentially contains Mg-trioctahedral layers. Muscovite is also available in a large single crystalline form.

Kaolinite is a 1:1 clay which can be frequently found in sedimentary rocks. Water molecules do not usually intercalate kaolinite. However, with an appropriate chemical treatment, it is possible to prepare water-intercalated kaolinites.[2] Below, we will discuss pressure effects on a water-intercalated kaolinite with a basal spacing of 8.6 Å, prepared from Georgia well-crystallized kaolinite (KGa-1, Source Clays Repository of the Clay Minerals Society). This monohydrate kaolinite is quite stable under ambient conditions.

HYDRATION STATES

Figure 1 shows results from isotherm water adsorption experiments on Na-vermiculite. The amount of water adsorbed per one gram of dry vermiculite at 301 K is plotted as a function of vapor pressure. Below p = ~3 Torr, the sample is in the zero water layer hydration state (WLHS) with a basal spacing of 9.8 Å. As the vapor pressure increases, the weight of the sample increases and reaches a plateau at p = ~6 Torr. This plateau corresponds to the one WLHS. The basal spacing is now 11.9 Å. Further increase in the vapor pressure causes a transition form the one WLHS to the two WLHS at p = ~15 Torr. Na-vermiculite in the

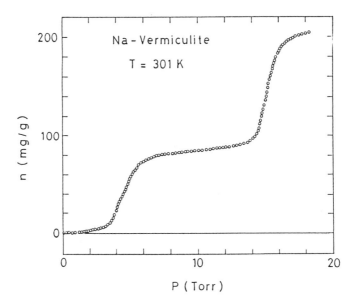

Figure 1. Isotherm water adsorption for Na-vermiculite. The amount of water adsorbed (in mg) per gram of anhydrous Na-vermiculite is plotted against water vapor pressure.

two WLHS yields 14.8 Å for the basal spacing. From the weight increase in the sample, the number of the water molecules per Na atom are estimated to be ~2.2 and ~6.2 for the one and two WLHS, respectively. These transitions in the hydration states are of first order, evidenced by a large hysteresis. By increasing the sample temperature, one observes increases in the transition pressures. It is emphasized that one can obtain a well-defined hydration state in vermiculite and may be able to study in detail the intercalation mechanism, various inter- and intralayer interactions (e.g. Coulombic, hydrogen bond, cation-water molecule, van der Waals and so forth.). Detailed discussion on the thermodynamics and phase diagram of vermiculite intercalates will be found elsewhere.[3]

MAGNETIC PROPERTIES

The interlamellar cations can also be exchanged with magnetic ions such as Ni^{2+}, Co^{2+}, Mn^{2+}, Cu^{2+}, Fe^{2+}, Fe^{3+} and so on. Since the interlayer distance (thus interlayer magnetic interactions) can be controlled by preparing different hydration states and the cations can be mixed with different magnetic or non-magnetic ions, one envisages that the magnetic vermiculite intercalates may potentially provide an exciting opportunity for studying low-dimensional magnetism.

Ni-vermiculite can be prepared in the zero, one and two WLHS with the basal spacing, 10.2, 12.2 and 14.8 Å, respectively.[4] Figure 2 shows an example of

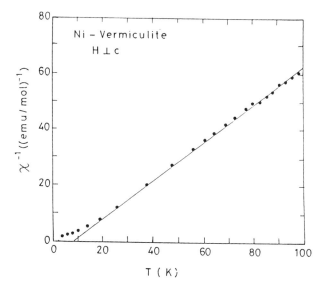

Figure 2. Reciprocal static susceptibility of Ni-vermiculite in the two WLHS as a function of temperature. The applied magnetic field is perpendicular to the c-axis.

static susceptibility data from Ni-vermiculite. The figure shows a Curie-Weiss behavior at high temperatures, but a deviation from it at lower temperatures. The static susceptibility data from Ni-vermiculite yield Curie-Weiss temperatures, $\Theta(\parallel c) = 12 \pm 3$ K and $\Theta(\perp c) = 10 \pm 2$ K for the two WLHS, and $\Theta(\parallel c) = 19 \pm 2$ K and $\Theta(\perp c) = 16 \pm 3$ K for the zero WLHS. These positive and orientationally dependent Curie-Weiss temperatures indicate an anisotropic ferromagmetic ordering. Note also that the Curie-Weiss temperatures increase when the basal spacing changes from 14.8 Å to 10.2 Å. The interlayer magnetic interaction becomes stronger in the zero WLHS than in the two WLHS. The effective magnetic moment was found to be ~ 3.3 μ_B, indicating that the Ni-ions exist as divalent.

Figure 3 shows AC susceptibility from Ni-vermiculite in the two WLHS with the magnetic field parallel to the c-axis as a function of temperature. We note two interesting features. One is that there are peaks at 2.7 and 2.2 K, indicating *two* transitions. The other is that the value of the AC susceptibility decreases as the temperature decreases (which is typically seen in spin glass systems.). Further investigation will clarify the nature of the transitions.

LATTICE DYNAMICS

Figures 4 and 5 show the phonon dispersion curves obtained from single-crystal muscovite by inelastic neutron scattering.[5] Previous data by Cebula *et al.*[6]

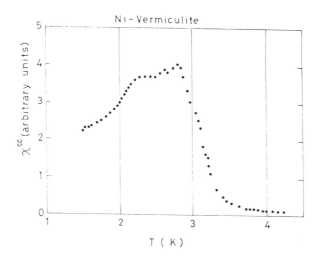

Figure 3. AC susceptibility of Ni-vermiculite in the two WLHS as a function of temperature. The applied magnetic field is along the c-axis.

are also shown in Fig. 4 for comparison. The slopes of the acoustic branches yield the bulk moduli, $K_{||}$ = 5.4 x 10^{11} dyne/cm^2 and $K_{\perp}$ = 7.8 x 10^{12} dyne/cm^2. As expected, the ratio $K_{\perp}/K_{||}$ (= ~14) indicates a highly anisotropic layered structure. This anisotropy can be also seen by the phonon curve for the in-plane TA mode with the polarization perpendicular to the plane, showing a characteristic $\omega^2 \sim Aq^2 + Bq^4$ behavior.[7] Note that the interlayer shearing mode which corresponds to the TO zone-center phonon in Fig. 4 has an energy of 3.6 mev (29 cm^{-1}). Because of weak interlayer interactions and a relatively large unit cell, the interlayer shearing frequencies are very low.

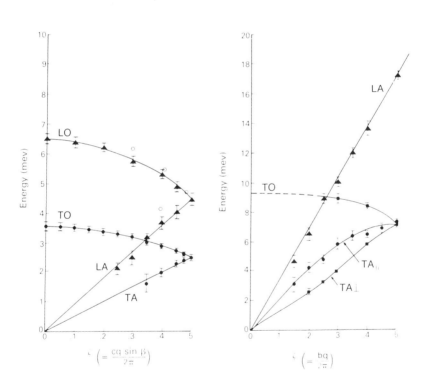

Figure 4. Phonon dispersion curves obtained from muscovite. The momentum transfer, $\bar{q}$ is parallel to the c-axis. The c-axis lattice constant, c = 20.0 Å, and β = 96°. Previous data taken by Cebula et al.[6] were also shown by ○ for comparison with our data.

Figure 5. Phonon dispersion curves obtained from muscovite. The momentum transfer, $\bar{q}$ is parallel to the b-axis. The b-axis lattice constant, b = 8.9 Å.

Figure 6 shows spectra taken from Na-vermiculite in the zero, one, and two WLHS by Raman scattering.[5,8] The Raman peak at 24 cm^{-1} in the anhydrous Na-vermiculite spectrum may be assigned to the interlayer shearing mode (ILSM) and corresponds to the TO zone-center mode mentioned above. From hydrous Na-vermiculite, no corresponding ILSM peak was observed. Because of softening in the interlayer interactions upon water intercalation, the ILSM frequency may become so low that the corresponding Raman peak will be buried under the laser stray light background. We also examined Sr-vermiculite and found that the ILSM frequency to be 21 cm^{-1} in the zero WLHS. No corresponding ILSM Raman peak was found in the hydrated states, either.

The ILSM frequency should be quite sensitive to the interlamellar cations, because the intralayer structure of the silicate layer is very rigid. The frequency

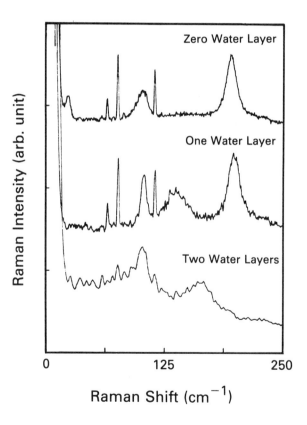

Figure 6. Raman spectra taken from Na-vermiculite in the zero, one and two WLHS. A 5145 Å Ar-ion laser beam was used. The resolusion-limited peaks at 66, 77 and 117 cm^{-1} are laser plasma lines.

difference between Na- and Sr-vermiculites may be understood in terms of the areal density of the interlayer cations. The number of Na^+ atoms per unit area in vermiculite is half that of Sr^{2+}. This implies that the interlayer shearing force constant of Na-vermiculite should be larger than that of Sr-vermiculite. (Note that the ionic radii of Na^+ and Sr^{2+} are similar.) The same sort of explanation may hold for the frequency difference between muscovite and Na-vermiculite, i.e. muscovite has more charge per unit area and thus a stronger interlayer interaction than Na-vermiculite.

The Raman peak at 196 cm^{-1} for the Na-vermiculite in the zero WLHS shows a drastic change as a function of the hydration state, as is seen in Fig. 6. This mode might correspond to the intralayer E_2 mode,[9] which involves the surface oxygen atoms in the tetrahedral sheets. This mode is thus expected to be sensitive to the interlamellar intercalant structure. When in the two WLHS, Na^+ ions occupy positions above the tetrahedral bases because of electrostatic interactions between the Na^+ ions and the substituted Al^{3+} ions in the Si^{4+} sites in the tetrahedral sheet. This two WLHS exhibits a broad peak at 164 cm^{-1} shown in Fig. 6. However, when the hydration state of Na-vermiculite becomes one-layered, the Na-cations can be situated over either the tetrahedral bases or the ditrigonal cavities. With this, the mode in the two WLHS becomes perturbed and splits into two modes at 137 and 199 cm^{-1}. Further hydration causes all the Na-cations to sit in the ditrigonal cavities. This yields a relatively high-frequency Raman peak at 196 cm^{-1}, since the intralayer oxygen bonds are stiffened by the presence of Na-atoms in the cavities.

PRESSURE EFFECTS

Figure 7 shows x-ray diffraction patterns taken from kaolinite samples immersed in various solvents. Each sample had been kept in a high pressure vessel with a solvent at 1 kbar for 65 hours, and was then examined by x-ray scattering when the sample was still wet with the solvent. Notice the drastic changes in the x-ray diffraction patterns from the kaolinite samples immersed in methanol, water and acetone. In the acetone case, the transformation of the 8.6-Å kaolinite to an acetone-intercalated kaolinite (11.2 Å) was complete. In the cases of water and methanol, the transformations were not complete, i.e. the x-ray patterns exhibited a mixture of the original 8.6-Å and new intercalated kaolinite Bragg peaks. The fraction of the new phases were found to increase with time. Once dried, all of the newly intercalated kaolinites exhibited the original 8.6 Å peak, indicating deintercalation. The intercalated states are metastable. We did not observe noticeable changes in the x-ray patterns of the water-intercalated kaolinite with ethanol, chloroform and dichloroform solvents from that of the original 8.6-Å kaolinite.

When the water-intercalated kaolinite was immersed in those solvents, no intercalation was observed in the time scale of a few hours. However, in the time

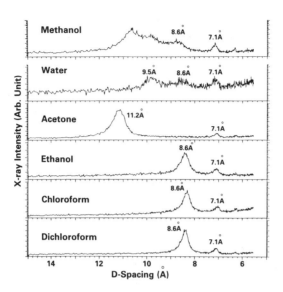

Figure 7. X-ray diffraction patterns from the kaolinite samples when they were still wet with the solvents. Each sample had been kept in a pressure vessel at ~1 kbar with the solvent. The CuKα x-ray line was used to obtain the patterns. The small peak at 7.1 Å originates from the original kaolinite.

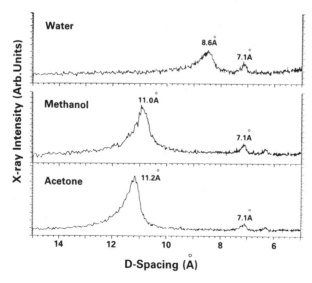

Figure 8. X-ray diffraction patterns from the kaolinite. Each sample had been in the solvent at ambient pressure for 15 days.

scale of days, we found diffusion controlled intercalation in some cases. Figure 8 shows x-ray diffraction patterns obtained from the kaolinite which had been in water, methanol and acetone. Methanol and acetone intercalated the kaolinite, and the basal spacings expanded to ~ 11.0 and 11.2 Å, respectively. Notice that the water-intercalated kaolinite (8.6 Å) *was not* intercalated with water in these experiments. We *did not* observe any intercalation in the kaolinite with ethanol, either.

In addition, high-pressure x-ray diffraction experiments were conducted on the water-intercalated kaolinite, using a diamond anvil cell.[10] Figure 9 plots pressure versus the basal spacing for the original water-intercalated kaolinite (8.6 Å) intercalated with methanol, ethanol and a 4:1 mixture of methanol and ethanol (this mixture of alcohol is commonly used as a high-pressure transmitting media because of the high freezing pressure). Note that the basal spacings of these kaolinites expanded because of the intercalation. Even ethanol was found to

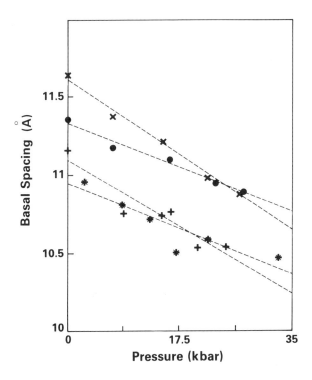

Figure 9. Basal spacing versus pressure. X: DMSO kaolinite in the 4:1 alcohol, O: 8.6 Å kaolinite in the 4:1 alcohol, +: 8.6 Å kaolinite in ethanol, and ✶: 8.6 Å kaolinite in methanol. The dotted straight lines show best least-square fits.

intercalate the kaolinite. (This implies that the critical pressure needed for the intercalation is somewhat above a few kbar.) Also is shown the data obtained from DMSO- (dimethylsulfoxide) intercalated kaolinite with the alcohol mixture as a pressure transmitting medium for comparison. We did not observe any observable change in the basal spacing of DMSO-kaolinite, implying that the mixture of the alcohols did not intercalate the DMSO-kaolinite.

It is concluded that pressure is needed for water and ethanol to intercalate the water-intercalated kaolinite (8.6 Å). Acetone and methanol intercalate the kaolinite diffusively with time. It is curious that ethanol and methanol behaved differently as described above.

CONCLUSION

We discussed experimental results from well-characterized clays. In vermiculite, one finds well-defined hydration states, which are functions of temperature and vapor pressure. X-ray, neutron and Raman scattering techniques were used to obtain information on the structure of vermiculite intercalates. It is now of great interest to investigate the dynamics of the transitions in the hydration states, e.g. how the domains of interlamellar water molecules form, grow and behave as a function of temperature, vapor pressure and time. One may also find interest in investigating the quasi-two dimensional magnetic properties of vermiculite intercalates, since one can prepare vermiculite intercalates with various magnetic ions (even a mixture) and can control the dimensionality by changing the hydration state.

Pressure is as important a thermodynamical parameter as temperature, but has not received much attention in clay research. It is of crucial importance for us who are interested in porous rocks to understand the properties of clays under *high pressure* and high temperature conditions. The high pressure experiments on kaolinite described above, for instance, imply that reservoir rocks in laboratory conditions may behave quite differently in borehole conditions.

It is hoped that that these fundamental experiments should lead to a detailed understanding of intercalation and other low-dimensional phenomena in clays and consequently porous rocks.

ACKNOWLEDGMENTS

This work was done in partial collaboration with D. R. Hines, W. A. Kamitakahara, K. Koga, S. Minomura, H. Nishihara, R. Raythatha, M. Suzuki and M. S. Whittingham. Useful discussions with M. Lipsicas, S. A. Solin and P. Wong are greatly appreciated.

REFERENCES

1. see, for instance, R. E. Grim, *Clay Mineralogy*, (McGraw-Hill, New York, 1968) and *Crystal Structures of Clay Minerals and Their X-ray Identification*, edited by G. W. Brindly and G. Brown (Mineralogical Society, London, 1980).

2. R. Raythatha and M. Lipsicas, Clay and Clay minerals, **33** 333 (1985).

3. M. Suzuki, N. Wada, D. R. Hines and M. S. Whittingham, to be published in MRS symposium proceedings, *Graphite Intercalation Compounds* ed. by M. S. Dresselhaus, G. Dresselhaus and S. A. Solin, (Materials Research Society, Pittsburg, 1986).

4. K. Koga, H. Nishihara, M. Suzuki and N. Wada, to be published in MRS symposium proceedings, *Graphite Intercalation Compounds* ed. by M. S. Dresselhaus, G. Dresselhaus and S. A. Solin, (Materials Research Society, Pittsburg, 1986).

5. N. Wada, W. A. Kamitakahara and M. Suzuki, to be published in MRS symposium proceedings, *Graphite Intercalation Compounds* ed. by M. S. Dresselhaus, G. Dresselhaus and S. A. Solin, (Materials Research Society, Pittsburg, 1986).

6. D. J. Cebula, M. C. Owen, C. Skinner, W. G. Stirling and R. K. Thomas, Clay Minerals, **17** 195 (1982).

7. H. Zabel, W. A. Kamitakahara and R. M. Nicklow, Phys. Rev. B **26** 5919 (1982).

8. N. Wada, in *Proceedings of the Tenth International Conference on Raman Spectroscopy* ed. by W. L. Peticolas and B. Hudson, (Univ. of Oregon, Eugene, 1986).

9. M. Ishii, T. Shimanouchi and M. Nakahira, Inorg. Chim. Acta **1** 387 (1967).

10. N. Wada, Phys. Rev. B, **24** 1065 (1981).

IV. Mechanical Properties

EXTENSIONS OF BIOT'S THEORY OF POROELASTICITY TO COMPLEX POROUS MEDIA

James G. Berryman and Lewis Thigpen
Lawrence Livermore National Laboratory
P. O. Box 808 L-156
Livermore, CA 94550

ABSTRACT

We summarize recent efforts to extend the theory of poroelasticity to semilinear and nonlinear elastic response, to partially saturated pores, to inhomogeneous frame materials, and to viscous losses due to localized flow effects. The prospects for a comprehensive theory of wave propagation in porous media are also discussed.

INTRODUCTION

In his original derivation, Biot constructed a theory of poroelasticity that was quite limited in scope.[1,2] He assumed linear, isotropic elastic response on the macroscopic scale for porous media composed of homogeneous frame materials and fully saturated pores. The principal attenuation mechanism of this theory was viscous attenuation due to shear induced during macroscopic flow of the single-phase fluid filling the pores. Even with these simplifications, the resulting theory has remained a scientific oddity for 30 years: (1) It is relatively hard to analyze the predictions of this theory[3] because it involves two coupled wave equations forming a system somewhat more complex than the equations of viscoelasticity[4] – which are nontrivial to analyze themselves! (2) The most startling predictions of the theory – such as the existence of a slow bulk compressional wave[1] or slow surface[5] and extensional waves[6] – are often very hard to verify in the laboratory.[7-11] (3) Even the physical interpretation of many of the coefficients in the equations remained unclear for 25 years,[8,12-14] and in some cases are still in dispute today.[15] It is therefore understandable that significant progress towards eliminating the many simplifying assumptions contained in the original work had not been made prior to the present decade. Indeed, why complicate a subject which is already so difficult?

The reason of course is "realism." For many of the geophysical applications of most interest, the pertinent geological materials are anisotropic and very heterogeneous, composed of multiple frame materials and multiple pore fluids. In some applications, the exciting waves are of large amplitude so that linear equations of motion are simply inadequate to describe the phenomena we want to study. Often we argue that the elementary theory should suffice to explain the gross behavior of such materials, justifying our approximations with the comparative simplicity and elegance of the resulting theory. If the theory is really successful at explaining the preponderance of experimental data, then of course our arguments are justified and it would appear to be of only academic interest to expend such effort as would be required to construct a truly comprehensive theory. On the other hand, since the theory to date has been unable to explain some of the most elementary experimental results for waves in geological materials, it seems to us essential to proceed with a program of research designed to produce a sophisticated theory capable of treating most of the complications encountered in practice.

Other researchers have also produced extensions of the elementary theory. Biot himself had generalized the theory to include anisotropic effects for dynamic problems[16] and nonlinear effects for quasistatic problems.[17] Various other authors have treated the generalization to partial saturation at very low frequencies in an intuitively appealing manner,[18-22] but without having any clear procedure for generalizing their results for higher frequencies. The form of the equations for the elastic coefficients when the frame material is composed of two or more constituents has been known for some time,[22] but no method for obtaining the required data has been suggested.

The authors have initiated a general program of research aimed at producing a comprehensive theory of dynamic poroelasticity. Although irreversible pore collapse[23] is important in some of our applications, we have neglected such effects initially in order to construct what is otherwise a quite general Lagrangian variational principle[24] for nonlinear and semilinear (reversible) deformations of dry and fluid saturated porous solids. This approach is very closely related to an Eulerian variational formulation by Drumheller and Bedford[25] for flow of complex mixtures of fluids and solids. We have shown that our theory reduces correctly to Biot's equations of poroelasticity[1] for small amplitude wave propagation and that it also reduces correctly to Biot's theory[17] of nonlinear and semilinear rheology for porous solids when the deformations are sufficiently slow. The resulting theory is a nontrivial generalization of Biot's ideas including explicit equations of motion for changes of solid and fluid density. Furthermore, if we assume that capillary pressure effects may be neglected, then the linear theory also shows that calculations on problems with only partially saturated pores may be reduced to computations of the same level of difficulty as those for fully saturated pores.[26] We expect the general theory to give a very good account of the behavior of wet porous materials during elastic deformations. Figure 1 provides a schematic illustration of the goals for this program, but of course much of this work has been done by others and much of it remains to be done.

In the presentation that follows, we will concentrate on three recent extensions of the theory of poroelasticity that tend to make the theory more realistic for applications to rocks. First, we show how the theory may be generalized to partially saturated porous media. Then, we use an effective medium approach to find estimates of the coefficients in the equations when the frame material is inhomogeneous. Finally, we analyze the attenuation of the fast compressional wave in heterogeneous media and show that the physically correct damping coefficient depends not on the global permeability, but on a simple spatial average of the local permeability.

SIMPLIFIED EQUATIONS FOR PARTIAL SATURATION

When the mechanical and thermodynamical processes set in motion by a deformation are reversible, an energy functional which includes all the important effects involved in the motion may be constructed. Equations of motion may then be found by an application of Hamilton's principle. Such variational methods based on energy functionals are well-known in continuum mechanics.[27] Thus, the only really new feature in the present context is the degree of complexity; porous earth may be composed of many types of solid constituents and the pore space may be filled with a mixture of water and air. Some irreversible effects may also be included in the variational method (e.g., losses of energy due to drag between constituents) when they may be analyzed in terms of a dissipation functional. Other irreversible effects such as those associated with collapse of the pore space lie outside the scope of the traditional variational approaches; the forms normally used for the energy functionals are quadratic with constant coefficients in the linear problems or simply positive definite polynomials with constant coefficients

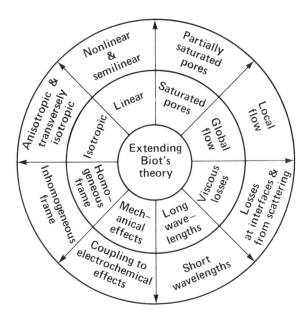

Fig. 1. The Expanding Microcosm of Poroelasticity – Schematic diagram of the many directions research must take to produce a comprehensive theory of wave propagation in porous media.

for nonlinear problems. During pore collapse, the usual assumptions about the form of the energy functionals are violated, so the usefulness of the variational method is questionable. However, if we restrict discussion to linear processes, the variational methods are entirely adequate.

Using these variational methods, we have shown[26] that the general equations of motion for linear elastic wave propagation through a porous medium containing both liquid and gas in the pores are given by

$$\Lambda_{(\xi)0}\ddot{\bar{\rho}}_{(\xi)} = -\frac{\partial E_{(\xi)}}{\partial \bar{\rho}_{(\xi)}} - \frac{\lambda_\phi}{\bar{\rho}^2_{(\xi)0}}, \qquad (1)$$

$$\rho_{(s)0}\ddot{u}_{(s)i} + \sum_{\gamma=g,l}\rho_{(s\gamma)0}(\ddot{u}_{(s)i} - \ddot{u}_{(\gamma)i}) = \left[\rho_{(s)0}\frac{\partial E_{(s)}}{\partial u_{(s)i,j}} + \lambda_\phi\phi_{(s)0}\delta_{ij}\right]_{,j} + d_{(s)i} + \rho_{(s)0}b_{(s)i}, \qquad (2)$$

and

$$\rho_{(\gamma)0}\ddot{u}_{(\gamma)i} + \sum_{\xi\neq\gamma}\rho_{(\gamma\xi)0}(\ddot{u}_{(\gamma)i} - \ddot{u}_{(\xi)i}) = \phi_{(\gamma)0}(\lambda_\phi)_{,i} + d_{(\gamma)i} + \rho_{(\gamma)0}b_{(\gamma)i} \qquad (3)$$

where $\gamma = g$ or l and $\xi = g, l,$ or s. The displacements are $u_{(\xi)i}$; the local densities (mass per unit volume of constituent) are $\bar{\rho}_{(\xi)}$; and the partial densities (mass per unit total volume) are $\rho_{(\xi)} = \phi_{(\xi)}\bar{\rho}_{(\xi)}$. The internal energies of these immiscible constituents are $E_{(\xi)}$. The induced mass coefficients are $\rho_{(s\gamma)0}$. The body forces are given by $b_{(\xi)i}$ and the drag forces by $d_{(\xi)i}$. The authors have shown elsewhere[28] that the drag forces may be written in the form $d_{(\gamma)i} = -\sum_\xi D_{(\gamma\xi)}(\dot{u}_{(\xi)i} - \dot{u}_{(s)i})$ where $D_{(\gamma\xi)}$ is a symmetric, positive semidefinite matrix whose matrix elements satisfy $\sum_\gamma D_{(\gamma\xi)} = 0$ for $\xi = g$ or l.

One major simplification that occurs in the equations for partial saturation follows from (1) and the approximation $\Lambda_{(\xi)0} = 0$. We find that

$$\lambda_\phi = -\bar{\rho}_{(\xi)0}^2 \frac{\partial E_{(\xi)}}{\partial \bar{\rho}_{(\xi)}} \equiv -p_{(\xi)} \tag{4}$$

where $p_{(\xi)}$ is the pressure for constituent ξ. Eq. (4) implies that all the pressures are equal – which is consistent with an assumption that capillary pressure effects are negligible. The subscript may subsequently be dropped from p. If $2e_{(\xi)ij} \equiv u_{(\xi)i,j} + u_{(\xi)j,i}$, then the first two strain invariants are defined by $I_{(\xi)1} = e_{(\xi)ii}$ and $I_{(\xi)2} = \frac{1}{2}[I_{(\xi)1}^2 - e_{(\xi)ij}e_{(\xi)ji}]$. The changes in density are defined by $\Delta\bar{\rho}_{(\xi)} = \bar{\rho}_{(\xi)} - \bar{\rho}_{(\xi)0}$. In terms of these invariants, the standard definitions of the internal energies are

$$\rho_{(s)0}E_{(s)} = \frac{1}{2}aI_{(s)1}^2 + bI_{(s)2} + cI_{(s)1}\Delta\bar{\rho}_{(s)} + \frac{1}{2}d\Delta\bar{\rho}_{(s)}^2 \tag{5}$$

and

$$\rho_{(\gamma)0}E_{(\gamma)} = \frac{1}{2}h_{(\gamma)}\Delta\bar{\rho}_{(\gamma)}^2 \tag{6}$$

for $\gamma = g, l$. Applying (4) to (5) and (6), we find

$$p = \frac{\bar{\rho}_{(s)0}}{\phi_{(s)0}}(cI_{(s)1} + d\Delta\bar{\rho}_{(s)})$$

$$= \frac{\bar{\rho}_{(l)0}}{\phi_{(l)0}}h_{(l)}\Delta\bar{\rho}_{(l)} \tag{7}$$

$$= \frac{\bar{\rho}_{(g)0}}{\phi_{(g)0}}h_{(g)}\Delta\bar{\rho}_{(g)}.$$

The coefficients in (5) have been shown elsewhere[24] to be related to known quantities: $a = \phi_{(s)0}K^*/(\sigma - \phi_{(f)0}) + \frac{4}{3}\mu^*$, $b = -2\mu^*$, $c = \phi_{(s)0}K^*/\bar{\rho}_{(s)0}(\sigma - \phi_{(f)0})$, and $d = [\phi_{(s)0}/\bar{\rho}_{(s)0}]^2 K_{(s)}/(\sigma - \phi_{(f)0})$ where $\sigma = 1 - K^*/K_{(s)}$. The bulk and shear moduli of the drained porous solid frame are K^* and μ^*. The bulk modulus of the (assumed) single constituent composing the microscopically homogeneous frame is $K_{(s)}$. If the solid frame is composed of two or more constituents, then these formulas must be modified. The coefficient $h_{(\gamma)}$ is related to the bulk modulus $K_{(\gamma)}$ of the γ-th fluid constituent by

$$h_{(\gamma)} = \frac{\phi_{(\gamma)0}}{\bar{\rho}_{(\gamma)0}^2}K_{(\gamma)}. \tag{8}$$

Now we define the linearized increment of fluid content for partial saturation to be

$$\zeta \equiv I_{(s)1} + \sum_\xi \frac{\phi_{(\xi)0}}{\bar{\rho}_{(\xi)0}}\Delta\bar{\rho}_{(\xi)}. \tag{9}$$

If only one fluid phase is present, (9) reduces to the exact result obtained previously by Berryman and Thigpen.[24] If more than one fluid phase is present, then we observe that by defining an effective total fluid density change according to

$$\frac{\phi_{(f)0}}{\bar{\rho}_{(f)0}}\Delta\bar{\rho}_{(f)} \equiv \frac{\phi_{(g)0}}{\bar{\rho}_{(g)0}}\Delta\bar{\rho}_{(g)} + \frac{\phi_{(l)0}}{\bar{\rho}_{(l)0}}\Delta\bar{\rho}_{(l)} \qquad (10)$$

with $\phi_{(f)0} \equiv \sum_\gamma \phi_{(\gamma)0}$ and we find that (9) reduces again to the exact result. Furthermore, applying (8), it is straight forward to show that (4) implies that

$$\frac{\Delta\bar{\rho}_{(\gamma)}}{\bar{\rho}_{(\gamma)0}} = \frac{p}{K_{(\gamma)}} \qquad (11)$$

for $\gamma = g$ or l. Substituting (11) into both sides of (10) shows that the effective bulk modulus of the multiphase fluid is given by

$$\frac{\phi_{(f)0}}{K_{(f)}} \equiv \sum_\gamma \frac{\phi_{(\gamma)0}}{K_{(\gamma)}} \qquad (12)$$

which is just the harmonic mean or Reuss average of the constituents' bulk moduli.

To check the consistency of our definition of ζ, we can show easily that

$$\zeta = \sum_\gamma \phi_{(\gamma)0}[I_{(s)1} - I_{(\gamma)1}]. \qquad (13)$$

If we define the average displacement of a fluid relative to the solid frame by

$$w_{(\gamma)i} = \phi_{(\gamma)0}[u_{(\gamma)i} - u_{(s)i}] \qquad (14)$$

for $\gamma = g$ or l and the total relative fluid displacement by

$$w_i = \sum_\gamma w_{(\gamma)i}, \qquad (15)$$

then (13) becomes

$$\zeta = -w_{i,i}. \qquad (16)$$

Eq. (16) reduces to the standard definition for full saturation when only one fluid saturates the pore space and is a natural generalization of this definition for partially saturated materials.

The total relative fluid displacement w_i defined by (15) is important in partial saturation problems not only because of the analogy just developed with the fully saturated problems, but also for convenience in applying boundary conditions in practical problems. We have shown previously[26] that uniqueness of the solutions to the equations (1) – (3) demands the specification of either p or the normal component of this same w_i on the boundaries of the porous material. Therefore, it proves most convenient to combine these equations so that $u_{(s)i}$ and w_i are the dependent variables. We will subsequently drop the subscript (s) on u_i since no confusion will arise and also define $e \equiv I_{(s)1}$. In addition, the zero subscripts on density and volume fraction may also be dropped in the remainder of the analysis.

To determine the relations among p, ζ, and e, substitute (11) and the first equation of (4) into (8) to eliminate $\Delta \bar{\rho}_{(\xi)}$ for all ξ. Using known identities and rearranging terms, we find easily that

$$p = M\zeta - Ce \qquad (17)$$

where the coefficients C and M are given by

$$C = \left\{ [(\sigma - \phi_{(g)} - \phi_{(l)})/K_{(s)} + \phi_{(g)}/K_{(g)} + \phi_{(l)}/K_{(l)}]/\sigma \right\}^{-1}, \qquad (18)$$

and

$$M = C/\sigma \qquad (19)$$

with

$$\sigma = 1 - K^*/K_{(s)}. \qquad (20)$$

Substituting (12) into (18) gives

$$C = \left\{ [(\sigma - \phi_{(f)0})/K_{(s)} + \phi_{(f)0}/K_{(f)}]/\sigma \right\}^{-1} \qquad (21)$$

which is the standard result for single-phase saturation.[29]

Next we suppose the body forces vanish and sum the equations (2) and (3) to obtain

$$\rho \ddot{u}_i + \bar{\rho}_{(g)} \ddot{w}_{(g)i} + \bar{\rho}_{(l)} \ddot{w}_{(l)i} = \left[\rho_{(s)} \frac{\partial E_{(s)}}{\partial u_{i,j}} - p\delta_{ij} \right]_{,j} \qquad (22)$$

where $\rho = \sum_\xi \rho_{(\xi)}$. Dividing (3) through by $\phi_{(f)0}$ and rearranging terms, we find

$$\bar{\rho}_{(g)} \ddot{u}_i + \frac{\alpha_{(g)} \bar{\rho}_{(g)}}{\phi_{(g)}} \ddot{w}_{(g)i} + \frac{D_{(gg)}}{\phi_{(g)}^2} \dot{w}_{(g)i} - \frac{\rho_{(gl)}}{\phi_{(g)}^2} \ddot{w}_{(l)i} = -p_{,i} \qquad (23)$$

and

$$\bar{\rho}_{(l)} \ddot{u}_i + \frac{\alpha_{(l)} \bar{\rho}_{(l)}}{\phi_{(l)}} \ddot{w}_{(l)i} + \frac{D_{(ll)}}{\phi_{(l)}^2} \dot{w}_{(l)i} - \frac{\rho_{(lg)}}{\phi_{(l)}^2} \ddot{w}_{(g)i} = -p_{,i}. \qquad (24)$$

In (23) and (24), $\alpha_{(g)}$ is the electrical tortuosity of the pore space occupied only by the gas, while $\alpha_{(l)}$ is the electrical tortuosity of the pore space occupied only by the liquid. Introducing a Fourier time dependence of the form $exp(-i\omega t)$ into (23) and (24), combining, rearranging terms, and keeping the same names for the transformed and untransformed variables, we have

$$-\omega^2 \begin{pmatrix} q_{(g)} & -r_{(g)} \\ -r_{(l)} & q_{(l)} \end{pmatrix} \begin{pmatrix} w_{(g)i} \\ w_{(l)i} \end{pmatrix} = \begin{pmatrix} -p_{,i} + \omega^2 \bar{\rho}_{(g)0} u_i \\ -p_{,i} + \omega^2 \bar{\rho}_{(l)0} u_i \end{pmatrix} \qquad (25)$$

where

$$\phi_{(\gamma)0}^2 q_{(\gamma)} \equiv \alpha_{(\gamma)} \rho_{(\gamma)0} + iD_{(\bar{\gamma}\gamma)}/\omega \qquad (26)$$

and

$$\phi_{(\gamma)0}^2 r_{(\gamma)} \equiv \rho_{(gl)0}. \qquad (27)$$

In (26), $\bar{\gamma} \neq \gamma$ so $\bar{\gamma} = l$ or g as $\gamma = g$ or l. Inverting the matrix in (25) and summing the results gives

$$-\omega^2[q_{(g)}q_{(l)} - r_{(g)}r_{(l)}]w_i = -(s_{(g)} + s_{(l)})p_{,i} + \omega^2[s_{(g)}\bar{\rho}_{(l)} + s_{(l)}\bar{\rho}_{(g)}]u_i \qquad (28)$$

where

$$s_{(\gamma)} = q_{(\gamma)} + r_{(\gamma)}. \qquad (29)$$

Using the expressions for $w_{(\gamma)i}$ from (25) again, we find

$$\sum_\gamma \bar{\rho}_{(\gamma)} w_{(\gamma)i} = \{[\bar{\rho}_{(g)}(q_{(l)} + r_{(g)}) + \bar{\rho}_{(l)}(q_{(g)} + r_{(l)})]p_{,i}$$
$$- \omega^2[q_{(l)}\bar{\rho}_{(g)}^2 + (r_{(g)} + r_{(l)})\bar{\rho}_{(g)}\bar{\rho}_{(l)} + q_{(g)}\bar{\rho}_{(l)}^2]u_i\}/\omega^2[q_{(g)}q_{(l)} - r_{(g)}r_{(l)}] \qquad (30)$$
$$= \frac{\bar{\rho}_{(g)}(q_{(l)} + r_{(g)}) + \bar{\rho}_{(l)}(q_{(g)} + r_{(l)})}{s_{(g)} + s_{(l)}} w_i - \frac{(\bar{\rho}_{(g)} - \bar{\rho}_{(l)})^2}{s_{(g)} + s_{(l)}} u_i$$

where $p_{,i}$ has been eliminated in the second step of (30) using (28).

The final form of these equations is found by substituting (30) into (22), using (18) in the result and also in (28), and finally rearranging terms. The equations then take the familiar form

$$\mu \nabla^2 \vec{u} + (H - \mu)\nabla e - C\nabla \zeta + \omega^2(\rho_{uu}\vec{u} + \rho_{uw}\vec{w}) = 0, \qquad (31)$$

$$C\nabla e - M\nabla \zeta + \omega^2(\rho_{wu}\vec{u} + \rho_{ww}\vec{w}) = 0, \qquad (32)$$

where the inertial coefficients are given by

$$\rho_{uu} = \rho - \frac{(\bar{\rho}_{(g)} - \bar{\rho}_{(l)})^2}{s_{(g)} + s_{(l)}}, \qquad (33)$$

$$\rho_{wu} = \frac{\bar{\rho}_{(g)}s_{(l)} + \bar{\rho}_{(l)}s_{(g)}}{s_{(g)} + s_{(l)}} = \rho_{uw} + \frac{(r_{(l)} - r_{(g)})(\bar{\rho}_{(g)} - \bar{\rho}_{(l)})}{s_{(g)} + s_{(l)}}, \qquad (34)$$

and

$$\rho_{ww} = \frac{q_{(g)}q_{(l)} - r_{(g)}r_{(l)}}{s_{(g)} + s_{(l)}}. \qquad (35)$$

The coefficient H is given by

$$H = K^* + \frac{4}{3}\mu^* + \sigma C \qquad (36)$$

while C and M are given by (18) and (19). Thus, we find the remarkable result that the equations of motion for partial saturation and for full saturation are the same – the only difference being that the inertial coefficients are more complex when the porous solid is only partially saturated.

BIOT'S THEORY OF POROELASTICITY

Now we will change notation somewhat and consider two porous media (i.e., host and inclusion) each of whose connected pore space is saturated with a single-phase viscous fluid. The fraction of the total volume occupied by the fluid is the void volume fraction or porosity ϕ, which is assumed to be uniform within a constituent but which may vary between the the host and inclusion. The bulk modulus and density of the fluid are K_f and ρ_f, respectively, in the host. The bulk and shear moduli of the drained porous frame for the host are K and μ. For now we assume the frame of the host is composed of a single constituent whose bulk and shear moduli and density are K_m, μ_m, and ρ_m. Corresponding parameters for the inclusion will be distinguished by adding a prime superscript. The frame moduli may be measured directly or they may be estimated using one of the many methods developed to estimate elastic constants of composites.[30]

For long-wavelength disturbances ($\lambda > h$, where h is a typical pore size) propagating through such a porous medium, we define average values of the (local) displacements in the solid and also in the saturating fluid. The average displacement vector for the solid frame is $\vec{u}$ while that for the pore fluids is $\vec{u}_f$. The average displacement of the fluid relative to the frame is $\vec{w} = \phi(\vec{u}_f - \vec{u})$. For small strains, the frame dilatation is

$$e = e_x + e_y + e_z = \nabla \cdot \vec{u}, \tag{37}$$

where e_x, e_y, e_z are the Cartesian strain components. Similarly, the average fluid dilatation is

$$e_f = \nabla \cdot \vec{u}_f \tag{38}$$

(e_f also includes flow terms as well as dilatation) and the increment of fluid content is defined by

$$\zeta = \nabla \cdot \vec{w} = \phi(e - e_f). \tag{39}$$

With these definitions, Biot[1,2,31] shows that the strain-energy functional for an isotropic, linear medium is a quadratic function of the strain invariants[32] $I_1 = e, I_2$, and of ζ having the form

$$2E = He^2 - 2Ce\zeta + M\zeta^2 - 4\mu I_2, \tag{40}$$

where

$$I_2 = e_y e_z + e_z e_x + e_x e_y - \frac{1}{4}(\gamma_x^2 + \gamma_y^2 + \gamma_z^2), \tag{41}$$

and $\gamma_x, \gamma_y, \gamma_z$ are the shear strain components. Our earlier definitions (5) and (6) for partial saturation are completely consistent[24,33] with these definitions.

With time dependence of the form $exp(-i\omega t)$, the Fourier transformed version of the coupled wave equations of poroelasticity in the presence of dissipation take the form

$$\mu \nabla^2 \vec{u} + (H - \mu)\nabla e - C\nabla \zeta + \omega^2(\rho \vec{u} + \rho_f \vec{w}) = 0, \tag{42}$$

$$C\nabla e - M\nabla \zeta + \omega^2(\rho_f \vec{u} + q\vec{w}) = 0, \tag{43}$$

where

$$\rho = \phi \rho_f + (1 - \phi)\rho_m \tag{44}$$

and

$$q = \rho_f [\alpha/\phi + iF(\xi)\eta/\kappa\omega]. \tag{45}$$

The kinematic viscosity of the liquid is η, the permeability of the porous frame is κ, and the dynamic viscosity factor[2] is given (for our present choice of sign for the frequency dependence) by

$$F(\xi) = \frac{1}{4}\xi T(\xi)/[1 + 2T(\xi)/i\xi], \tag{46}$$

where

$$T(\xi) = \frac{ber'(\xi) - ibei'(\xi)}{ber(\xi) - ibei(\xi)} \tag{47}$$

and

$$\xi = (\omega h^2/\eta)^{\frac{1}{2}}. \tag{48}$$

The functions $ber(\xi)$ and $bei(\xi)$ are the real and imaginary parts of the Kelvin function. The dynamic parameter h is a characteristic length generally associated with (and comparable in magnitude to) the steady-flow hydraulic radius. The electrical tortuosity α is a pure number related to the frame inertia which has been measured recently[14] for porous glass bead samples and has also been estimated theoretically.[8,12] The electrical tortuosity α and the fluid flow tortuosity τ are related by $\alpha = \tau^2 = \phi F$, where F is the electrical formation factor.

The coefficients H, C, and M are given by[22,29]

$$H = K + \frac{4}{3}\mu + \sigma C, \tag{49}$$

$$C = \{[(\sigma - \phi)/K_m + \phi/K_f]/\sigma\}^{-1}, \tag{50}$$

$$M = C/\sigma, \tag{51}$$

where

$$\sigma = 1 - K/K_m. \tag{52}$$

To decouple the wave equations (42) and (43) into Helmholtz equations for three modes of propagation, we note that the displacements $\vec{u}$ and $\vec{w}$ can be decomposed as

$$\vec{u} = \nabla \Upsilon + \nabla \times \vec{\beta}, \quad \vec{w} = \nabla \psi + \nabla \times \vec{\chi}, \tag{53}$$

where Υ, ψ are scalar potentials and $\vec{\beta}, \vec{\chi}$ are vector potentials. Substituting (53) into Biot's equations (42) and (43), we find they are satisfied if two pairs of equations hold:

$$(\nabla^2 + k_s^2)\vec{\beta} = 0, \quad \vec{\chi} = -\Gamma_s \vec{\beta}, \tag{54}$$

where $\Gamma_s = \rho_f/q$ and

$$(\nabla^2 + k_\pm^2)A_\pm = 0. \tag{55}$$

In this notation, the subscripts $+, -,$ and s refer respectively to the fast and slow compressional waves and the shear wave. The wave vectors in (54) and (55) are defined by

$$k_s^2 = \omega^2(\rho - \rho_f \Gamma_s)\mu \tag{56}$$

and

$$k_\pm^2 = (\omega^2/2\Delta)(b + f \mp [(b-f)^2 + 4cd]^{\frac{1}{2}}), \tag{57}$$

where
$$b = \rho M - \rho_f C, \; c = \rho_f M - qC, \; d = \rho_f H - \rho C, \; f = qH - \rho_f C, \qquad (58)$$
with
$$\Delta = MH - C^2. \qquad (59)$$
The linear combination of scalar potentials has been chosen to be
$$A_\pm = \Gamma_\pm \Upsilon + \psi, \qquad (60)$$
where
$$\Gamma_\pm = d/[(k_\pm \Delta/\omega^2)^2 - b] = [(k_\pm \Delta/\omega^2)^2 - f]/c. \qquad (61)$$
With the identification (61), the decoupling is complete.

Since (54) and (55) are valid for any choice of coordinate system, they may be applied to boundary value problems with arbitrary symmetry. Biot's theory has therefore been applied to the scattering of elastic waves from a spherical inhomogeneity by Berryman.[34] The results of that calculation will be summarized in the next section.

SCATTERING FROM A SPHERICAL INHOMOGENEITY

The full analysis of scattering from a spherical inhomogeneity in a fluid-saturated porous medium is quite tedious. Fortunately, much of this work has already been done[34] and we may therefore merely quote the pertinent results here.

Let the spherical inhomogeneity have radius a. For the moment, we will place no restrictions on the properties of the inhomogeneous region. Thus the frame bulk and shear moduli, the grain bulk modulus, the density, the porosity, and the permeability of a solid inclusion may all be different from those of the host. Furthermore, the bulk modulus, density, and viscosity of the fluid in an inhomogeneous region may also all be different from those of the host fluid. Suppose now that a plane fast compressional wave is generated at a free surface far from the inclusion. Then, if the incident fast compressional wave has the form
$$\vec{u} = \hat{z} \frac{A_0}{ik_+} exp\, i(k_+ z - \omega t), \qquad (62)$$
the radial component of the scattered compressional wave contains both fast and slow parts in the far field and is given by
$$u_{1r} = (ik_+)^{-1} exp\, i(k_+ r - \omega t)/k_+ r [B_0^{(+)} - B_1^{(+)} \cos\theta - B_2^{(+)}(3\cos 2\theta + 1)/4] \\ - (ik_-)^{-1} exp\, i(k_- r - \omega t)/k_- r [B_0^{(-)} - B_1^{(-)} \cos\theta - B_2^{(-)}(3\cos 2\theta + 1)/4]. \qquad (63)$$
Then, with the definitions $\kappa_\pm = k_\pm a$ and $\kappa_s = k_s a$ and with no restrictions on the materials, we find that
$$B_0^{(-)} = \frac{i\kappa_-^3 A_0}{3M'(\Gamma_+ - \Gamma_-)(K' + \frac{4}{3}\mu)} \Big[(C - M\Gamma_-)(K' + \frac{4}{3}\mu) \\ - (C' - M'\Gamma_-)(K + \frac{4}{3}\mu) + (C - M\Gamma_-)(C' - M'\Gamma_-)\Big(\frac{C'}{M'} - \frac{C}{M}\Big) \Big], \qquad (64)$$
and
$$B_0^{(+)} = \frac{\kappa_+^3 A_0}{3i} \frac{[K' - K + (C - M\Gamma_-)(C'/M' - C/M)]}{K' + \frac{4}{3}\mu} + (\kappa_+/\kappa_-)^3 B_0^{(-)}. \qquad (65)$$

Expansions of the other coefficients in the small parameter $\epsilon = C/K$ have been given in the reference.[34] However, for the present application, only the first two coefficients are needed and these happen to be the only ones known exactly at present. Of course, the full scattered wave also contains transverse components of the compressional wave, relative fluid/solid displacement, and mode converted shear waves. However, the scattering coefficients for these contributions are linearly dependent on the the coefficients in (63) and therefore contain no new information. It is sufficient then to base our discussion on the expression (63).

As an elementary check on our analysis, we should first consider the limit in which the porosity ϕ vanishes. Then the fluid effects disappear from the equations and only the first line of (63) survives. Furthermore, it is not difficult to check[34] that the coefficients $B_n^{(+)}$ for $n = 0, 1, 2$ reduce to the well-known results for scattering from a spherical elastic inclusion in an infinite elastic medium.[30] For example,

$$B_0^{(+)} = -i\kappa_+^3 A_0 (K' - K)/(3K' + 4\mu) \tag{66}$$

in this limit as expected.

These results have a multitude of potential uses. One straightforward application is the calculation of energy losses from elastic wave scattering by randomly distributed particles. A second important application is to use these results as the basis for an effective medium approximation for the effective constants of complex porous media. The second application is the one we will address in the next section.

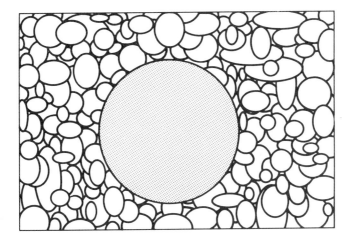

Fig. 2. A spherical inhomogeneity in a porous medium could be caused by local variations in fluid content, grain composition, porosity, permeability, etc.

MICROSCOPIC HETEROGENEITY

As we have noted previously, the equations of poroelasticity have several significant limitations. For example, these equations were derived with an explicit long-wavelength (low-frequency) assumption and also with strong implicit assumptions of homogeneity and isotropy on the macroscopic scale. Another restriction assumes that the pore fluid is uniform and that it fully saturates the pore space. For the present application, we will assume that a single fluid saturates all the pore space for host as well as inclusion and the scattering is caused by microscopic heterogeneity in the solid properties.

Before deriving our main results, consider the problem of the porous frame without a saturating fluid (or with a highly compressible saturating gas). Then, since we take $C = M = \rho_f = 0$ in this limit, each term of Eq. (50) vanishes identically and the fluid dependent terms of Eq. (49) also vanish, leaving only the terms for the elastic behavior of the porous frame remaining. Since no slow wave can propagate under these circumstances, the second line of Eq. (63) disappears and only the fast wave terms contribute to the scattering. This limit is formally equivalent to the problem of elastic wave scattering from a spherical inclusion which has been treated in detail previously (see Ref. 30 and other references therein). The effective medium approximation requires the weighted average of the single-scattering results to vanish. This approach simulates the physical requirement that the forward scattering should vanish at infinity if the impedance of the "effective medium" has been well matched to that of the composite. The resulting condition is that the volume weighted average of each of the $B_n^{(+)}$'s for $n = 0-2$ must vanish. Using the convention that the effective constants for the composite porous medium are distinguished by an asterisk, the formulas for the effective bulk (K^*) and shear(μ^*) moduli for the drained porous frame of a microscopically heterogeneous medium are

$$\frac{1}{K^* + \frac{4}{3}\mu^*} = \left\langle \frac{1}{K(\vec{x}) + \frac{4}{3}\mu^*} \right\rangle \qquad (67)$$

and

$$\frac{1}{\mu^* + F^*} = \left\langle \frac{1}{\mu(\vec{x}) + F^*} \right\rangle \qquad (68)$$

where

$$F = (\mu/6)(9K + 8\mu)/(K + 2\mu). \qquad (69)$$

The spatial($\vec{x}$) average is denoted by $\langle \cdot \rangle$. The remaining constant to be determined is the corresponding effective density which is just given by the average density.[34] Eq. (67) follows easily from the volume average of (66), while Eq. (68) follows similarly from the volume average of B_2^+. Note that the equations for K^* and μ^* are coupled and therefore must be solved iteratively (i.e., self-consistently). Although the form of the equations (67) and (68) is identical to that obtained for elastic composites, it is important to recognize that the results can be quite different since the local constants $K(\vec{x})$ and $\mu(\vec{x})$ appearing in the formulas are frame moduli of the constituent spheres of drained porous material, not (necessarily) the moduli of the individual material grains. Of course, since the formula reduces correctly in the absence of porosity to the corresponding result for the purely elastic limit, the user of Eqs. (67) and (68) has some discretion about conceptually lumping grains together to form a porous frame or treating them as isolated elastic inclusions. For purposes of modelling complex aggregates of grains typical of earth materials, this freedom of choice appears to be a real advantage.

Now we will restrict discussion to the very low frequency limit where

$$\Gamma_+ = H/C \tag{70}$$

and

$$\Gamma_- = 0. \tag{71}$$

With these restrictions, the relevant scattering coefficients reduce to

$$B_0^{(-)} = \frac{i\kappa_-^3 C A_0}{3HM'(K' + \frac{4}{3}\mu)} \left[C(K' + \frac{4}{3}\mu + \sigma'C') - C'(K + \frac{4}{3}\mu + \sigma C) \right], \tag{72}$$

and

$$B_0^{(+)} = \frac{\kappa_+^3 A_0}{3i} \frac{[K' - K + (\sigma' - \sigma)C]}{K' + \frac{4}{3}\mu} + (\kappa_+/\kappa_-)^3 B_0^{(-)}. \tag{73}$$

The resulting conditions on the effective constants are

$$\left\langle \frac{C^*\left[K(\vec{x}) + \frac{4}{3}\mu^* + \sigma(\vec{x})C(\vec{x})\right] - C(\vec{x})\left[K^* + \frac{4}{3}\mu^* + \sigma^*C^*\right]}{M(\vec{x})[K(\vec{x}) + \frac{4}{3}\mu^*]} \right\rangle = 0 \tag{74}$$

and

$$\left\langle \frac{K(\vec{x}) - K^* + [\sigma(\vec{x}) - \sigma^*]C^*}{K(\vec{x}) + \frac{4}{3}\mu^*} \right\rangle = 0. \tag{75}$$

Recall that the averages in (74) and (75), as elsewhere in this paper, refer to spatial averages over (possibly) porous constituents of the overall porous aggregate. The limitations on the assumed geometry of the resulting aggregate have been discussed previously.[35] Note that (74) and (75) depend on the effective medium frame moduli K^* and μ^* determined by (67) and (68). The new constants determined by (74) and (75) are C^* and σ^*. The expressions for C^* and σ^* are coupled as written but may be uncoupled after some algebra. The final expressions for these constants are

$$C^* = \sigma^* \Big/ \left[\left\langle \frac{1}{M(\vec{x})} \right\rangle + \left\langle \frac{\sigma^2(\vec{x}) - (\sigma^*)^2}{K(\vec{x}) + \frac{4}{3}\mu^*} \right\rangle \right] \tag{76}$$

and

$$\sigma^* = \left\langle \frac{\sigma(\vec{x})}{K(\vec{x}) + \frac{4}{3}\mu^*} \right\rangle \Big/ \left\langle \frac{1}{K(\vec{x}) + \frac{4}{3}\mu^*} \right\rangle. \tag{77}$$

Notice that both constants are determined explicitly by the formulas, in contrast to the frame moduli K^* and μ^* which are determined only implicitly by (67) and (68). It has also been shown[35] that (76) and (77) are completely consistent with all known constraints[22,29] on the form of these coefficients.

The same idea used to derive (76) and (77) was also used to show[36] that the speed of waves propagating through a mixture of liquid and gas in the low frequency limit is given by Wood's formula[37] as expected.[20,21]

LOCAL-FLOW BIOT THEORY

A convincing demonstration has been given[38] that, if we assume global fluid-flow effects dominate the viscous dissipation, Biot's theory of poroelasticity cannot explain the observed magnitude of wave attenuation in partially saturated rocks. Since the same theory explains the wave speeds quite well, it is reasonable to suppose that a small change in the theory may be adequate to repair this flaw. Many possible explanations are possible of course, but within the context of Biot's theory the simplest postulate is to suppose that local – rather than global – fluid-flow effects dominate the dissipation.[15] We will distinguish two related issues in this section which are summarized in the following questions: (1) Does the physics of wave propagation require that the value of the permeability κ appearing in the Biot's equations should be that for global flow or for local flow? Then, if we can show that the value should be that for local flow, (2) does this change in the interpretation make enough difference so that the theory can explain the correct magnitude for the attenuation?

To address the first question, we explore the consequences of assuming that Biot's theory should be applied at the local flow level rather than at the global flow level. This assessment is easily done by examining the dispersion relations. When the Fourier time dependence is $e^{-i\omega t}$ with angular frequency ω sufficiently low, Biot's theory predicts [see Eq. (57)] the dispersion relations for the fast (+) and slow (−) compressional modes in any homogeneous porous material to be

$$k_+^2 \simeq \frac{\omega^2}{v_+^2}[1 + i\omega \frac{\rho_f^2}{\rho q_0}(1 - v_0^2/v_+^2)^2] \tag{78}$$

and

$$k_-^2 \simeq \frac{i\omega q_0 H}{MH - C^2} \tag{79}$$

where

$$v_+^2 = H/\rho, v_0^2 = C/\rho_f, q_0 = \rho_f \eta/\kappa. \tag{80}$$

The fraction of the total volume occupied by the fluid is the void volume fraction or porosity ϕ, which is assumed to be uniform. The bulk modulus and density of the fluid are K_f and ρ_f. The bulk and shear moduli of the drained porous frame are K and μ. For simplicity we assume the frame is composed of a single constituent whose bulk and shear moduli and density are K_m, μ_m, and ρ_m. Then the coefficients H, C, and M are given by (49)-(52). The overall density is

$$\rho = \phi\rho_f + (1 - \phi)\rho_m. \tag{81}$$

The kinematic viscosity of the fluid is η and the permeability of the porous frame is κ.

Defining the quality factor for the fast compressional wave Q_+ by

$$k_+^2 = \frac{\omega^2}{v_+^2}[1 + i/Q_+], \tag{82}$$

we find that Q_+ is given by

$$1/Q_+ = \omega \frac{\kappa \rho_f}{\eta \rho}(1 - v_0^2/v_+^2)^2. \tag{83}$$

Since $1/Q_+$ is proportional to the permeability, the attenuation is therefore greatest in regions of high permeability. Thus, we might say that the regions of high permeability control the attenuation.

In the very low frequency limit, the slow compressional mode is known to reduce to Darcy flow with slowly changing magnitude and direction as the driving potential gradient oscillates sinusoidally.[39] Now consider a layered porous material (whose constants depend only on depth z) with constituents having identical physical constants except for the permeability κ which varies widely from layer to layer but which has a constant value κ_n within the n-th layer (lying in the range $z_{n-1} \leq z \leq z_n$ with $z_0 = 0$). Thus, the permeability is a piece-wise constant function of z. The thickness of the n-th layer is given by $l_n = z_n - z_{n-1}$. If we impose a potential gradient along the z-direction in such a layered material, it is well-known that the effective permeability for fluid flow is found by taking the harmonic mean of the constituent permeabilities, *i.e.*,

$$1/\kappa_f = \frac{1}{L}\sum_n \frac{l_n}{\kappa_n}, \tag{84}$$

where the total sample length L is given by the sum of the layer thicknesses

$$L = \sum_n l_n. \tag{85}$$

From (84), we can conclude that the regions of lowest permeability dominate the effective overall permeability for fluid flow through a porous layered medium. Thus, we might say that the regions of low permeability control the fluid flow – at least for this special choice of geometry.

The apparent attenuation of a fast compressional mode at normal incidence on such a structure has two distinct components: (1) Reflection and mode conversion at layer interfaces will have a tendency to degrade the fast wave, but this effect will be quite small at low frequencies for the model structure we are considering. (2) The attenuation within a layer is determined by the quality factor for that layer, as shown by Eq. (82). Assuming the attenuation is small enough, we may approximate (82) within any layer by $k_+(z) = \frac{\omega}{v_+}[1 + i/2Q_+(z)]$, where the functions $k_+(z)$ and $Q_+(z)$ take the piece-wise constant values appropriate for the depth argument z. Neglecting the small effects of reflection and mode conversion, the behavior of the fast compressional wave at normal incidence is then easily seen to be given by

$$A_+ exp[i\int_0^z dz k_+(z) - i\omega t] \simeq A_+ exp[i\frac{\omega}{v_+}z - i\omega t - \frac{\omega}{v_+}\int_0^z dz \frac{1}{2Q_+(z)}], \tag{86}$$

where A_+ is the amplitude of the wave at $z = 0$. In writing (86), we have used the piece-wise constant property of the functions. The integral in the exponent is given by

$$\int_0^z dz \frac{1}{2Q_+(z)} = \frac{\omega \rho_f}{2\eta\rho}(1 - v_0^2/v_+^2)^2 \int_0^z dz \kappa(z). \tag{87}$$

At the $z = L$ boundary of the material, we have

$$\int_0^L dz \kappa(z) = \sum_n l_n \kappa_n. \tag{88}$$

If the layering is periodic with period much less than either z or L or if it is statistically homogeneous on this length scale, then we may approximate the integral in the exponent of (86) using (87) and

$$\int_0^z dz \kappa(z) \simeq \kappa_a z, \tag{89}$$

where the effective permeability for attenuation measurements is given by the mean

$$\kappa_a = \frac{1}{L} \sum_n l_n \kappa_n. \tag{90}$$

It is well-known that the mean is always greater than or equal to the harmonic mean of any function; thus,

$$\kappa_f \leq \kappa_a. \tag{91}$$

The answer to our first question is therefore that the physics of wave propagation dictates that local-flow effects must dominate the attenuation of the fast compressional wave. The necessity of this conclusion is nicely illustrated in Figure 3. Suppose that a fast compressional wave is incident on a layered material with alternating permeable and impermeable layers. If the impermeable layers are very thin and have an acoustic impedance closely matching that of the permeable layers, their presence has a negligible effect on the propagating fast wave. The viscous attenuation of the fast wave occurs solely in the permeable layers and magnitude of that attenuation is completely determined by the permeability of these layers. By contrast, the global permeability of this material in the direction normal to the layering vanishes identically. If this zero value were used in our predictions, the magnitude of the attenuation would be grossly under estimated. Although this choice of geometry is extreme, it clearly shows that errors in estimates of attenuation will arise if the value of permeability for global flow is used.

Now, can the theory predict the correct magnitude for the attenuation even with this change in the interpretation of the permeability factor? To use our conclusions in a predictive mode – *i.e.*, to predict the wave attenuation from measurements of permeability – we need some independent means of measuring the local permeability distribution. Normal laboratory flow experiments will not suffice, because they necessarily measure the global permeability. One promising method of estimating the local permeability uses image processing techniques to measure pertinent statistical properties of rock topology from pictures of cross sections.[40] This approach is still under development and we will not attempt to describe it in detail here.

Another approach, which is ultimately much less satisfactory than the image processing method but much easier to use at present, is to suppose that we can obtain reasonable estimates of the local permeability κ_L from the known values of the global permeability κ_G, the tortuosity $\tau = (\phi F)^{\frac{1}{2}}$, and the porosity ϕ. To do so requires a formula, so we will use a form of the Kozeny-Carman relation derived recently by Walsh and Brace.[41] For tubes of arbitrary ellipsoidal (major and minor axes a,b) cross-section the effective permeability of straight sections of such tubes is given by $\kappa = (\pi/4A)[a^3 b^3/(a^2 + b^2)]$. The porosity for an ellipsoidal tube is $\phi = \pi a b/A$ and the specific surface area is well approximated by $s \simeq 2\pi[(a^2 + b^2)/2]^{\frac{1}{2}}/A$. Then, a Kozeny-Carman relation satisfied by κ, ϕ, and s can be shown to be

$$\kappa = \frac{1}{2} \frac{\phi^3}{s^2} \tag{92}$$

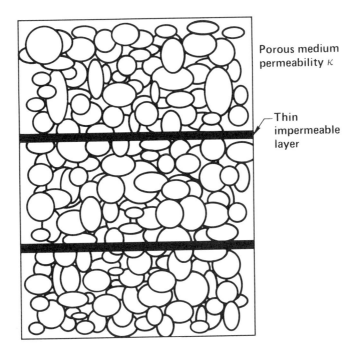

Fig. 3. Illustration of a simple experiment to prove that the attenuation of the fast wave depends on the local – not the global – value of permeability κ. A fast wave incident normal to the impermeable partitions will experience a small but finite attenuation even though the global permeability in this direction vanishes identically.

for the effective permeability of a single tube oriented along the pressure gradient. If the tube is at an angle θ to the this gradient, then it is not hard to show[41] that

$$\kappa = \frac{1}{2}\frac{\phi^3}{s^2\tau^2}, \qquad (93)$$

where $\tau = 1/cos\theta$. If we suppose that (92) and (93) are fairly representative of the material of interest, then (92) describes the maximum local permeability κ_L and (93) the effective global permeability κ_G. We then conclude that

$$\kappa_L = \tau^2 \kappa_G = \phi F \kappa_G. \qquad (94)$$

The tortuosity τ has been measured for many sandstones; the values for samples studied by Simmons et al.[42] lie in the range $1.5 \leq \tau \leq 5$, with most values $\tau \simeq 2$. To obtain estimates of attenuation close to experiment,[20] we need to increase the value of permeability used in Mochizuki's calculations[38] by a factor of $\tau^2 \simeq 10$. This requirement implies a tortuosity of $\tau \simeq 3$, which is clearly well within the established experimental bounds. This argument provides strong evidence for the plausibility of a local-flow explanation of the observed discrepancies. However, a completely satisfying demonstration must await the collection of the required data on local-flow permeability.

One unfortunate consequence of the observation that local permeability controls attenuation is that measured attenuation in wet rocks cannot be used as a diagnostic of the global fluid-flow permeability. Nevertheless, since the mean of the local permeabilities will always be greater than the true fluid-flow permeability regardless of the actual spatial distribution of the constituent κ's, the effective permeability obtained from attenuation measurements can be used to provide an upper bound on the desired global permeability.

DISCUSSION

What then are the prospects for a comprehensive theory of poroelasticity? It appears likely at this point that we will soon have a completely satisfactory linear theory of bulk waves including effects of partial saturation and inhomogeneous frame materials. A satisfactory nonlinear theory of bulk waves including effects of fracture, plastic flow, and pore collapse is at a more elementary stage, but is still likely to be achieved in the next decade. At present it appears that the problems most likely to cause real trouble are those involving surface waves rather than the bulk waves. Surface waves depend critically on the nature of the equations of motion near interfaces. Using the standard boundary conditions of poroelasticity,[26,43] it has been shown that a slow surface wave[5] or slow extensional wave[6] is expected only when a closed-pore boundary condition applies at the porous surface. Yet, the experimental data to date seems to show that such slow surface waves[10] do in fact propagate when the open-pore boundary condition applies. It is possible that the presence of a thin damage region close to the surface has a major effect on the conclusions of the theory regarding the propagation of the surface waves. However, it could also be that these experiments are pointing out still another subtle deficiency of the equations we use to describe wave propagation in porous media.

ACKNOWLEDGEMENTS

We thank S. C. Blair, B. P. Bonner, R. C. Y. Chin, and M. J. Miksis for very helpful discussions. Work performed under the auspices of the U. S. Department of Energy by the Lawrence Livermore National Laboratory under contract No. W-7405-ENG-48 and supported specifically by the Nuclear Test Containment Program and by the Office of Basic Energy Sciences.

REFERENCES

1. M. A. Biot, "Theory of propagation of elastic waves in a fluid-saturated porous solid. I. Low-frequency range," *J. Acoust. Soc. Am.* **28**, 168-178 (1956).

2. M. A. Biot, "Theory of propagation of elastic waves in a fluid-saturated porous solid. II. Higher frequency range," *J. Acoust. Soc. Am.* **28**, 179-191 (1956).

3. R. Burridge and C. A. Vargas, "The fundamental solution in dynamic poroelasticity," *Geophys. J. Roy. Astr. Soc.* **58**, 61-90 (1979).

4. R. C. Y. Chin, "Wave propagation in viscoelastic media," in *Physics of the Earth's Interior*, Proceedings of the Enrico Fermi Summer School 1979, Course LXXVIII (North-Holland, Amsterdam, 1980), pp. 213-246.

5. S. Feng and D. L. Johnson, "High-frequency acoustic properties of a fluid/porous solid interface. I. New surface mode," *J. Acoust. Soc. Am.* **74**, 906-914 (1983).

6. J. G. Berryman, "Dispersion of extensional waves in fluid-saturated porous cylinders at ultrasonic frequencies," *J. Acoust. Soc. Am.* **74**, 1805-1812 (1983).

7. T. J. Plona, "Observation of a second bulk compressional wave in a porous medium at ultrasonic frequencies," *Appl. Phys. Lett.* **36**, 259-261 (1980).

8. J. G. Berryman, "Confirmation of Biot's theory," *Appl. Phys. Lett.* **37**, 382-384 (1980).

9. D. Salin and W. Schön, "Acoustics of water saturated packed glass spheres," *J. Phys. Lett.* **42**, 477-480 (1981).

10. M. J. Mayes, P. B. Nagy, L. Adler, B. P. Bonner, and R. Streit, "Excitation of surface waves of different modes at fluid-porous solid interface," *J. Acoust. Soc. Am.* **79**, 249-252 (1986).

11. R. Lakes, H. S. Yoon, and J. L. Katz, "Slow compressional wave propagation in wet human and bovine cortical bone," *Science* **220**, 513-515 (1983); "Ultrasonic wave propagation and attenuation in wet bone," *J. Biomed. Engng.* **8**, 143-148 (1986).

12. R. J. S. Brown, "Connection between formation factor for electrical resistivity and fluid-solid coupling factor in Biot's equations for acoustic waves in fluid-filled porous media," *Geophysics* **45**, 1269-1275 (1980).

13. D. L. Johnson, "Equivalence between fourth sound in liquid He II at low temperatures and the Biot slow wave in consolidated porous media," *Appl. Phys. Lett.* **37**, 1065-1067 (1980).

14. D. L. Johnson, T. J. Plona, C. Scala, F. Pasierb, and H. Kojima, "Tortuosity and acoustic slow waves," *Phys. Rev. Lett.* **49**, 1840-1844 (1982).

15. J. G. Berryman, "Elastic wave attenuation in rocks containing fluids," *Appl. Phys. Lett.* **49**, 552-554 (1986).

16. M. A. Biot, "Generalized theory of acoustic propagation in porous dissipative media," *J. Acoust. Soc. Am.* **34**, 1254-1264 (1962).

17. M. A. Biot, "Nonlinear and semilinear rheology of porous solids," *J. Geophys. Res.* **78**, 4924-4937 (1973).

18. S. N. Domenico, "Effects of water saturation of sand reservoirs encased in shales," *Geophysics* **29**, 759-769 (1974).

19. S. N. Domenico, "Elastic properties of unconsolidated sand reservoirs," *Geophysics* **41**, 882-894 (1977).

20. W. F. Murphy, III, "Effects of partial water saturation on attenuation in Massilon sandstone and Vycor porous glass," *J. Acoust. Soc. Am.* **71**, 1458-1468 (1982).

21. W. F. Murphy, III, "Acoustic measures of partial gas saturation in tight sandstones," *J. Geophys. Res* **89**, 11549-11559 (1984).

22. R. J. S. Brown and J. Korringa, "On the dependence of the elastic properties of a porous rock on the compressibility of the pore fluid," *Geophysics* **40**, 608-616 (1975).

23. J. F. Schatz, "Models of inelastic volume deformation of porous geologic materials," in *The Effects of Voids on Material Deformation*, AMD – Vol. 16, edited by S. C. Cowin and M. M. Carroll (American Society of Mechanical Engineers, New York, 1976), pp. 141-170.

24. J. G. Berryman and L. Thigpen, "Nonlinear and semilinear dynamic poroelasticity

with microstructure," *J. Mech. Phys. Solids* **33**, 97-116 (1985).
25. D. S. Drumheller and A. Bedford, "A thermomechanical theory for reacting immiscible mixtures," *Arch. Rational Mech. Anal.* **73**, 257-284 (1980).
26. J. G. Berryman and L. Thigpen, "Linear dynamic poroelasticity with microstructure for partially saturated porous solids," *ASME J. Appl. Mech.* **52**, 345-350 (1985).
27. A. Bedford, *Hamilton's Principle in Continuum Mechanics*, Research Notes in Mathematics # 139 (Pitman, Boston, 1985).
28. L. Thigpen and J. G. Berryman, "Mechanics of porous elastic materials containing multiphase fluid," *Int. J. Eng. Sci.* **23**, 1203-1214 (1985).
29. F. Gassmann, "Über die elastizität poröser medien," *Veirteljahrsschrift der Naturforschenden Gesellschaft in Zürich* **96**, 1-23 (1951).
30. J. G. Berryman, "Long-wavelength propagation in composite elastic media I. Spherical inclusions," *J. Acoust. Soc. Am.* **68**, 1809-1819 (1980).
31. M. A. Biot, "Mechanics of deformation and acoustic propagation in porous media," *J. Appl. Phys.* **33**, 1482-1498 (1962).
32. A. E. H. Love, *A Treatise on the Mathematical Theory of Elasticity* (Dover, New York, 1944), pp. 43, 62, 102.
33. A. Bedford and D. S. Drumheller, "A variational theory of porous media," *Int. J. Solids Structures* **15**, 967-980 (1979).
34. J. G. Berryman, "Scattering by a spherical inhomogeneity in a fluid-saturated porous medium," *J. Math. Phys.* **26**, 1408-1419 (1985).
35. J. G. Berryman, "Effective medium approximation for elastic properties of porous solids with microscopic heterogeneity," *J. Appl. Phys.* **59**, 1136-1140 (1986).
36. J. G. Berryman and L. Thigpen, "Effective constants for wave propagation through partially saturated porous media," *Appl. Phys. Lett.* **46**, 722-724 (1985).
37. A. W. Wood, *A Textbook of Sound* (Bell, London, 1957), p. 360.
38. S. Mochizuki, "Attenuation in partially saturated rocks," *J. Geophys. Res.* **87**, 8598-8604 (1982).
39. D. L. Johnson, J. Koplik, and R. Dashen, "Theory of dynamic permeability and tortuosity in fluid-saturated porous media," *Bull. Am. Phys. Soc.* **31**, 576 (1986).
40. J. G. Berryman and S. C. Blair, "Use of digital image analysis to estimate fluid permeability of porous materials: Application of two-point correlation functions," *J. Appl. Phys.* **60**, 1930-1938 (1986).
41. J. B. Walsh and W. F. Brace, "The effect of pressure on porosity and the transport properties of rock," *J. Geophys. Res.* **89**, 9425-9431 (1984).
42. G. Simmons, R. Wilkens, L. Caruso, T. Wissler, and F. Miller, "Physical properties and microstructures of a set of sandstones," *Annual Report to the Schlumberger-Doll Research Center*, 1 January 1982, p. VI-16; ibid., 1 January 1983, p. VI-16.
43. H. Deresiewicz and R. Skalak, "On uniqueness in dynamic poroelasticity," *Bull. Seismol. Soc. Am.* **53**, 783-788 (1963).

POROSITY AND THE BRITTLE-DUCTILE TRANSITION IN SEDIMENTARY ROCKS

John M. Logan
Center for Tectonophysics and Department of Geophysics
Texas A&M University, College Station, TX 77843

ABSTRACT

The lower brittle-ductile transition, from macroscopic fracture to homogeneous cataclastic flow, although widely present in sedimentary rocks is poorly understood. It occurs when the shear stress necessary to activate dislocation slip exceeds that necessary for distributed microfracturing. The latter is facilitated by moderate to high porosities characteristic of some sedimentary rocks. This transition is common in quartz-rich clastics. A decrease of ultimate strength with increasing porosity is linear over the brittle and ductile regimes but follows a power law across the brittle-ductile transition. This larger effect is found to present at porosities of about 6-8% at confining pressures to 3.5 Kb. Sandstones with a simulated fracture show a transition to ductile behavior where the mechanism change is one from frictional sliding to homogeneous cataclastic flow. This sliding-failure envelope, contains different shear stresses as the frictional surface changes with displacement. It coalesces with that of the intact material in the ductile field. This transition is not abrupt but occurs over range of normal stress.

INTRODUCTION

One of the earliest experimental studies of the mechanical behavior of rocks by Frank Dawson Adams about 1901 demonstrated qualitatively that marble became ductile under sufficiently high confining pressure [1]. These experiments aided in resolving the controversy of whether rocks actually flowed or only appeared to do so. One side argued that they deformed only by brittle fracture which could subsequently heal while the other maintained that microscopic slip occurred promoting macroscopically ductile behavior. Quantitative documentation of the transition from macroscopically brittle to ductile behavior is given by T. von Karmen in 1911 in a classic sequence of experiments on marble and sandstone [2]. Using an experimental design which remains contemporary, he obtained differential stress--axial strain curves under a constant rate of shortening as a function of increasing confining pressure (Figure 1). He also noted that the ultimate and yield strengths increased with increasing confining pressure. With refreshing insight he described the macroscopic character of the deformed specimens. Brittle behavior is characterized by macroscopic fracture at failure, ductile by macroscopically homogeneous flow, and transitional the change from macro- to microscopic failure. Today, to our knowledge all rocks undergo such a transition which is promoted by increasing effective mean stress, temperature, and time of deformation.

What makes rocks interesting is that they are not all created equal. The necessary conditions for a brittle-ductile transition varies with composition, as do yield and ultimate strengths for the same conditions (Figure 2). The the differences in relative strength and ductility result from the conditions necessary to change the deformation mechanisms in rocks (Table I). In particular crystal plastic mechanism depend upon the structure of the crystal lattice which in turn is a function of the mineral composition. As common rocks have a great compositional variation the resulting mechanical spectrum is equally large. For instance the shear stress necessary to activate slip in halite, calcite, dolomite, and quartz varies by orders of magnitude [3,4]. Thus the concept of relative ductility is significant in explaining and predicting the behavior of rocks, influencing hydrofracture containment, salt, shale and granitic diaperic growth, location of thrust fault horizons and many other processes.

THE INFLUENCE OF POROSITY

Despite its early recognition, detailed understanding of the transition remains obscure. For instance, why is quartzite significantly stronger and more brittle than sandstones (Figure 2) and yet both have essentially the same composition--quartz?. Grain boundaries are generally are not as well defined in quartzites as they are in effect quartz cemented and some sandstones do contain some clay and calcite that would degrade their strength and increase the ductility. But many clean sandstones that are almost 100% quartz show strengths that are lower by factors of 10 than quartzites (Table II). Why? This question is intriguing because marble, which like quartzite is a metamorphic product, is not mechanically different than limestone its sedimentary counterpart (Table II).

First let us look at the brittle-ductile transition of these materials in more detail. Stress-strain curves showing the transition from brittle to ductile are similar in shape and amount of strain (e.g. marble and sandstone, Figure 1). Noticeable, of course, are the higher conditions necessary for quartzite to undergo the transition. Macroscopic observations show that all are similar to sandstone (Figure 3). Brittle failure is by through-going shear or extension fractures, which may vary in geometry, width and angle with the maximum compressive stress, but show general similarity regardless of rock type. The transition is complex and not well understood presently but shear bands are one diagnostic feature (Figure 4) [5]. It is marked by great sensitivity to subtle changes in the rock and boundary conditions; deformation may be concentrated at one end of the specimen (Figure 3), or shear fractures exist with shear bands. Ductile flow is accommodated at the microscopic level so that strain is unrecognizable at the macroscopic level without the aid of markers. It is at the microscopic level that the differences between sandstone and the other three rocks are evident.

In the ductile field, limestone, marble and quartzite all deform by crystal plastic mechanisms; dislocation glide or creep (Table I). These are thermally activated mechanisms, but for calcite at room temperature the critical resolved shear stress is only about 100 bars so that merely an increase in confining pressure activates them [6]. In quartz the temperatures are 800-900° C and confining pressures in excess of 5 Kb for these mechanisms to be activated [4]. This change to crystal plasticity has been referred to as the "upper or high temperature " transition (or Type A [7], or transition of first kind [8] and it is one commonly referred to in discussing this topic.

In contrast, the sandstone deforms by cataclastic flow-- homogeneous microfracturing associated with rigid body rotation and translation of the fragments and a general reduction of the grain size and porosity (Figure 5). The resulting rock framework is fundamentally different from those that deform by crystal plasticity and their other physical properties are accordingly changed by the difference in mechanisms. In that crystal plastic mechanisms are considered to be independent of normal stress, although some evidence suggests that this is not strictly true [9, 10]), it is expected that once the ductile state is dominated by dislocation slip and creep, the yield stress should no longer increase with confining pressure. The ultimate strength may increase due to strain hardening, but the yield stress remains relatively stable. This is observed in limestone, marble and quartzite [11,12,13,4]. In contrast, fracture and frictional sliding, the basis of cataclasis, are normal-stress dependent so the yield stress is found to increase with confining pressure in the ductile regime. [14, 15].

This "lower transition" (Type B [7] or transitions of the second kind [8]) is diagnostic of clastic sedimentary rocks which are quartz rich. It is found in non-porous quartzites but requires temperatures between 700-800°C [16]. Quartzites that contain about 7% porosity are reported to show this transition to cataclastic flow but at 6 Kb confining pressure at room temperature [17]. It is also a common mechanism in crustal fault zones [18]. But a textural feature of sedimentary rocks--moderate to high porosity--increases the environmental range of cataclastic behavior.

Consider the mechanical implications of medium to high porosity in sandstones. The pore space limits the grain-to- grain contacts, reducing the area over which the deforming loads are transmitted, producing stress risers at the contacts, resulting in fracturing at low boundary stresses. This effect has been modeled photoelastically and corroborated in experiments on quartz grains [19]. Once fracturing occurs, pieces rotate and translate into the pore space, producing a general comminution of the grains. Cataclasis, and corresponding ductility becomes more viable as porosity increases. While a decrease in ultimate strength with increasing porosity has been noted [20] what has been understood is the extent to which variations in porosity of the same composition can modify strength and ductility.

Recent experiments on Nugget sandstone document the effect of porosity changes on both strength and ductility. A series of triaxial tests were conducted at a shortening rate of 10^{-4}/sec., confining pressures to 3.5 Kb on specimens with porosities up to 15.1% . Nugget sandstone is about 98% quartz with trace amounts of calcite, clay and feldspar.

The ultimate strength decreases with increasing porosity (Figure 6) . At .345 Kb it is lowered by about 26%, while at 3.5 Kb it is about 47% less. So as an initial approximation, best fit linear curves are passed through the data, however the slopes are quite different.

If one plots changes in ultimate strength as a function of confining pressure, the changes are not linear except at the lowest porosities (Figure 7). At about 7.2% the curves become logarithmic with the slopes decreasing with further increases in porosity. Clearly the largest change occurs at the low to intermediate porosities, which is contrary to the model based upon decreasing grain contact with increasing porosity.

This paradox is resolved by looking at the failure strength with respect to the behavior of the rock in the brittle-ductile spectrum (Table III). Each experiment is evaluated as being brittle, transitional or ductile using criteria of stress-strain curve shape, strain to failure, and macroscopic deformation mechanisms. The low porosity specimens remain brittle over the confining pressure range, but those of 7.2 % porosity transitional at the higher pressures. Sandstones of 13.8 and 15.1% porosity become ductile at higher pressures. Referring the data to a traditional Mohr plot (Figure 8) where the points are calculated for the ultimate strength of each test these changes are reflected in the failure envelopes. The brittle specimens with porosities of 4.9 and 6.5% have linear envelopes. At higher porosities, as the sandstone becomes transitional and ultimately ductile, the slopes of the logarithmic functions decrease (Table IV). The greater sensitivity of ultimate strength to lower porosity is associated more directly with the failure mechanism in the brittle field--macroscopic shear fracture. This sensitivity decreases as the mechanisms shift more to microfracture. Thus, the cumulative effect of pervasive microfracturing and frictional slip when the rock becomes ductile, even though involving a much larger volume of the rock than macroscopic shearing, is to produce a much smaller normal stress dependence. Replotting the data of Figure 6 with this insight of physical process, the initial linear approximations for the specimens undergoing brittle to ductile transitions can be replaced with power-law functions (Figure 9). Interestingly, the largest changes in ultimate strength occur at the intermediate porosities--6.5 to 7.2%--that is where these rock begin to be transitional in behavior. Similar effects would be expected at higher pressures for sandstones of lower porosities, as they become transitional in behavior.

Diagnostic of the transition are associated volumetric changes. Dilatancy prior to macroscopic failure has been well documented [21] and is characteristic of the brittle field. With the transition to ductile behavior, it is suppressed until only compaction occurs with homogeneous cataclastic flow. This pore-volume reduction continues with increased deformation [22]. The absence of dilation is diagnostic of ductile behavior. A final observation is that even though the sandstone becomes ductile, a normal stress dependence persists reflecting the frictional dependence. More work needs to be done to investigate this dependence at higher pressures.

In summary, increasing porosity significantly reduces the ultimate strength and enhances the transition to ductile behavior. The latter occurs at much lower pressures in the rocks of higher porosity than in those of lower values. The decrease of ultimate strength with increasing porosity is most dramatic as the material changes from brittle to transitional behavior, with smaller changes occurring with ductile cataclastic flow.

FRACTURED ROCK AND THE BRITTLE-DUCTILE TRANSITION

Most rocks are fractured in their present state in the Upper Crust, therefore it is pertinent to ask what is the effect of these fractures on the brittle-ductile transition? It has been shown that as the rock becomes transitional, frictional sliding becomes restricted and eventually is totally inhibited as the material becomes ductile [23]. The processes involved in the transition from sliding to homogeneous flow, however, remain obscure.

Some insights can be gained by analyzing data on Berea sandstone which has a porosity of about 17%. Detailed experiments by W. Jamison and L. Teufel [24] on intact specimens deformed to span the brittle-ductile transition allow a failure envelope to be drawn (Figure 10). If fracture occurs, the shear stress-normal stress values for frictional sliding along the fracture can also be determined.

These data are compared with that acquired from experiments in which the specimen starts with a sawcut at 35° to the cylinder axis, and with the maximum compressive stress across the boundaries of the specimen . At the low confining pressures, shortening is entirely by frictional sliding along the saw cut. As the pressure is increased, sliding on the sawcut is inhibited after some displacement, and eventually a fracture forms with subsequent shortening by sliding along the fracture. At the highest pressures, no sliding along the sawcut occurs, with shortening accommodated by homogeneous cataclasis. That is, the specimen deforms ductilely, without any influence of the sawcut; the latter is almost indistinguishable following the experiment.

As expected the envelope for the intact specimen becomes nonlinear as the material becomes transitional (Figure 10), and shows normal stress dependence even in the ductile field. If the specimens fail by fracturing, which occurs at normal stresses up to about 1.5 Kb, the shear and normal stress drop to lower values, as displacement occurs along the fracture (Figure 10). With the sawcut-specimens, the shear stress necessary for sliding is below the failure envelop for the intact rock, as expected. Subsequent sliding either upon the sawcut or on a new fracture occurs at continually decreasing stress levels. Note for instance the levels at 2% and 7% shortening in specimens with sawcuts (Figure 10). The difference between the ultimate strength of the intact rock and that necessary to initiate sliding on the sawcut becomes increasingly smaller as the sandstone becomes transitional in its behavior. Finally when the sandstone is ductile, there is no longer any difference. Interestingly, the intersection of these two envelopes is not abrupt but the curves gradually merge to coincide. Although similar experiments have not yet been done for Nugget sandstone, the convergence should be similar.

Comparing the transition to ductile behavior where cataclastic flow is the mechanism to that where crystal plasticity occurs shows different idealized Mohr envelopes (Figure 11b). It has been found that the yield stress for crystal plasticity is temperature dependent but lacks normal-stress dependence [11]. Additionally, the intersection of the envelope for frictional sliding with that for intact rock should be relatively abrupt. For sandstones and other materials deforming by cataclastic flow (Figure 11a), the diagnostic feature is the persistence of normal stress dependence even when the rock becomes ductile. Additionally, there is no evidence of temperature dependence to date. The envelope for sliding is not a single line but a zone due to changing conditions along the surface with displacement, and the possibility of multiple sliding surfaces of different orientations. This envelope coalesces with that for the intact rock as the ductile field is entered.

An implication of these results is that the state of stress in the rock can change significantly even once fracture has occurred, particularly within the brittle field. This depends upon the character of the frictional surface and could result in widely different magnitudes within a rock mass, although the stress orientation would be relatively constant.

DISCUSSION

The decrease in strength with increasing porosity and the significance of the transition from brittle to ductile illustrate the importance of the changing physical mechanisms of failure. This is also apparent in the change from frictional sliding to homogeneous cataclastic flow. In both cases, increasing porosity decreases the brittle field, in the intact rock and that containing a simulated fracture. This change from localized failure, to localized, but distributed shearing in shear bands, and finally to distributed cataclastic flow is now well recognized. The theoretical basis of these changes remain obscure, however despite work on the concept of material stability [25,26,27]. Questions of stabilizing fracture propagation and concurrent enlargement of the fractured volume remain important. Of equal significance are the conditions resulting in shear bands where fracturing is stabilized but still localized (Figure 4). In that cataclasis, as a deformation mechanism, is largely restricted to rocks but is important in rock mechanics problems of the Upper Crust, these questions deserve future experimental and theoretical considerations.

ACKNOWLEDGEMENTS

The work on Nugget sandstone was done with encouragement of R. A. Nelson of AMOCO Producton Co., Tulsa and the work was generously supported by AMOCO Production Co.

REFERENCES

1. F. D. Adams, J. Geol., 18, 489 (1910).
2. T. von Karman, Zeits. . Ver. Dtsch. Ing. 55, 1749 (1911).
3. M. Friedman, Prod. 1st Intern. Cong. on Rock Mech., Lisbon, Portugal, III, 182 (1967).
4. H. C. Heard and N. L. Carter, Am. J. Sci., 266, 1 (1968).
5. M. Friedman and J. M. Logan, Geol. Soc. Am. Bull., 84, 1465 (1973).
6. M. Friedman, Jour. Geol., 71, 12 (1963).
7. K. Mogi, Tect., 13, 541 (1972).
8. S. A. F. Murrell, Gephy., J. Roy. Astr. Soc., 14, 81 (1967).
9. J. M. Edmond and M. Paterson, Int. J. Rock. Mech. Min. Sci., 9, 161 (1972).
10. T. Shimamoto and J. M. Logan, Am. Geoph. Un. Mon. 37, 49 (1986).
11. H. C. Heard, Geol. Soc. Am., Mem. 79, 193 (1960).
12. T. L. Blanton, Ph.D. Diss., Texas A&M Univ. p. 67 (1976).
13. H. C. Heard and C. B. Raleigh, Geol. Soc. Am. Bull., 83, 935 (1972).
14. J. W. Handin, Am. Assoc. Pet. Geol., 47, 717 (1963).
15. J. M. Logan, unpub. data.
16. G. Hirth and J. Tullis, Trans. Am. Geoph. Union, 67, 1186 (1986).
17. J. H. Benghazi and E. H. Rutter, Geol. Rundschau, 72, 493 (1983)
18. J. M. Logan, Am. Geoph. Un. Mon. 24, 121, (1981).
19. J. J. Gallagher, M. Friedman, J. Handin and G. Sowers, Tect., 21, 203, (1974).
20. D. E. Dunn, L. T., LaFountain and R. E. Jackson, Jour. Geoph. Res., 78, 2403 (1973).
21. C. H. Scholz, Jour. Geoph. Res., 73, 1417 (1968).
22. T. Shimamoto and J. M. Logan, Jour. Geoph. Res., 86, 2902 (1981).
23. J. M. Logan, Pure and Appl. Geop., 116, 773 (1978).
24. W. R. Jamison and L. W. Teufel, 20th U. S. Sym. Rock Mech., 163 (1979).
25. D. C. Drucker, J. Mec., 3, 235, (1964).
26. J. W. Rudnicki and J. R. Rice, J. Mech. Phys. Solids, 23, 371 (1975).
27. J. W. Rudnicki, J. Geoph. Res., 82, 844 (1977).

27. W. W. Krech, F. A. Henderson and K. E. Hjelmstad, U. S. Bur. Mines, RI 7865 (1974).
28. J. C. Jaeger and N. G. W. Cook, Fundamentals of Rock Mechanics (Methuen, N. Y. (1969) p. 513.
29. J. W. Handin and R. V. Hager, Bull. Am. Assic. Pet. Geol., 41, 1, (1958).

Table I. Common Deformation Mechanisms in Rocks

MACROSCOPIC FRACTURE	Shear or Extension Normal Stress Dependent
CATACLASIS	Microfracture and Rigid-Body Movements; Normal Stress Dependent
DISLOCATION GLIDE	Twin and Translation Gliding Temperature and Time Dependent
DISLOCATION CREEP	Dynamic Recrystallization Temperature and Time Dependent
DIFFUSION	Pressure Solution Grain-Boundary Diffusion (Coble Creep) Vacancy Diffusion (Nabarro-Herring Creep) Grain-Size Dependent

Table II. Comparison of Average Unconfined Tensile (T_o) and Compressive (C_o) Strengths

ROCK	C_o (Kb)	T_o (Kb)
Carrara Marble	.890	.068
Yule Marble	.390	
Salem Limestone	.440	.052
Holston Limestone	1.18	.010
Chesire Quartzite	4.60	.280
Sioux Quartzite	5.05	.010
Nugget Sandstone	1.25	
Berea Sandstone	.460	.010
Gosford Sandstone	0.50	.035

Data [27, 28].

Table III. Summary of failure mode in Nugget sandstone*.

Porosity (%)	Confining Pressure (Kb)					
	.345	.690	1.4	2.0	2.7	3.5
4.9	B-26/35°	B-15				
6.5	B-26			B-31		B-35
7.2	B-30	B-26	B-27	B-31	T-38	T-41
13.8	B-36	B-34		T-33		T
15.1	B-34	B-21		T	T	D

* Failure mode: B-brittle, T-transitional, D-ductile followed by fracture angle with respect to direction of maximum principle stress.

Table IV. Values of a and b derived from best fit curves of Figure 8.

POROSITY (%)	a	b
4.9*	-3.63	4.75
6.5**	0.950	0.661
7.2	1.423	0.609
13.8	1.223	0.557
15.1	1.189	0.518

* Curve of the form: $Y = aX + b + error$
** This and subsequent curves of the form: $Y = (a)X^b + error$

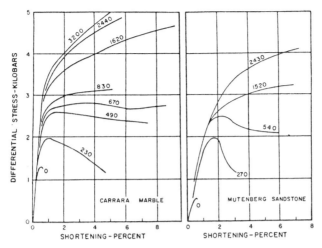

Fig. 1. Stress-strain showing brittle ductile transition as a function of confining pressure in bars. Modified from von Karmen [2].

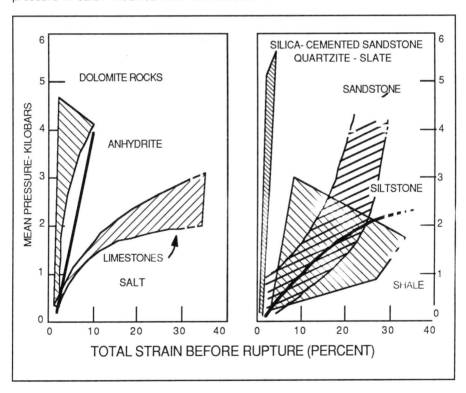

Fig. 2. Mean pressure versus total strain before rupture as a measure of ductility for the rock types shown. Data from Handin [29].

Fig. 3. Photograph of specimens showing transition from brittle to ductile, left to right respectively. The transition is as a function of increasing confining pressure. Specimens are about 2.5 cm. in diameter.

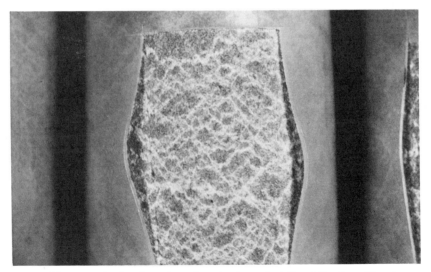

Fig. 4. Photograph showing shear bands characteristic of the transition to cataclastic flow. The rock is sandstone and specimen is about 10 cm. in diameter.

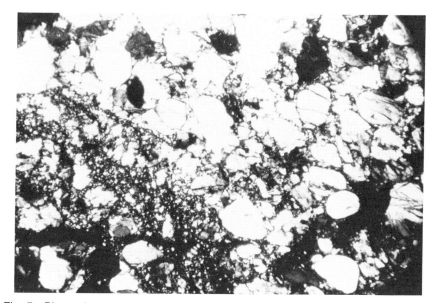

Fig. 5. Photomicrograph showing a sandstone which has deformed by cataclastic flow. Note microfractures within quartz grains. The field is about 5 mm across.

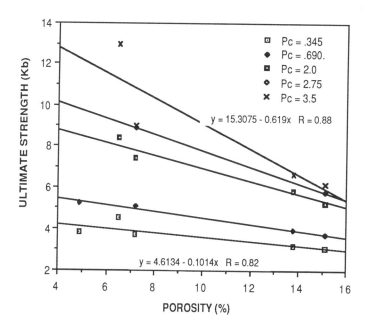

Fig. 6. Ultimate strength versus porosity for Nugget sandstone. Curves are for confining pressures indicated in kilobars. Equations for the upper and lower curves are shown.

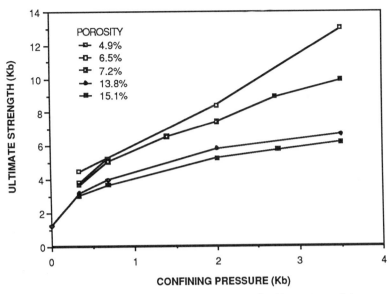

Fig. 7. Ultimate strength versus confining pressure for Nugget sandstone as a function of the porosities shown.

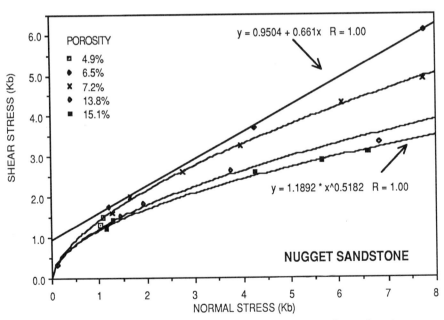

Fig. 8. Mohr diagram of Nugget sandstone. Points taken at failure of each experiment. Curves are for porosities shown. Equations for upper and lower curves as shown.

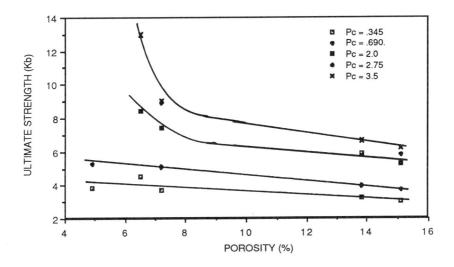

Fig. 9. Ultimate strength versus porosity for Nugget sandstone. Data the same as Fig. 6. Curves hand-drawn as best fit through data. Data points for confining pressures shown in kilobars.

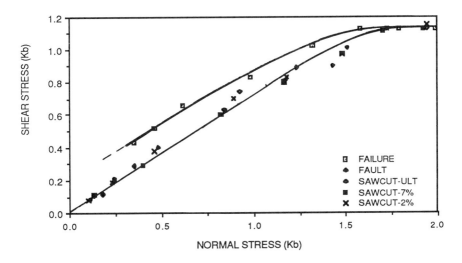

Fig. 10. Mohr plots for Berea sandstone. Data normalized from Jamison and Teufel [24]. Data points are for intact material where failure of the intact specimen shown as "Failure" and sliding along the created fracture at failure shown as "Fault". Data from specimens with a sawcut at 35° to the cylinder axis shown by "Sawcut". Sliding at the maximum differential stress designated as "Ult", after 2% sliding by "2%" and after 7% by "7%". See text for discussion. Curves passed through "Failure" and "Sawcut-7%" are hand-drawn best fit.

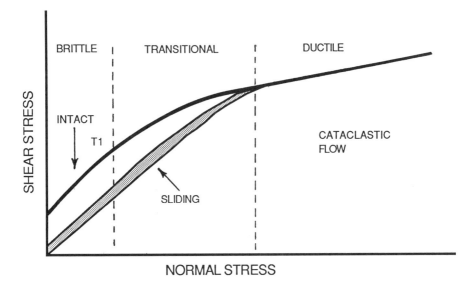

a.

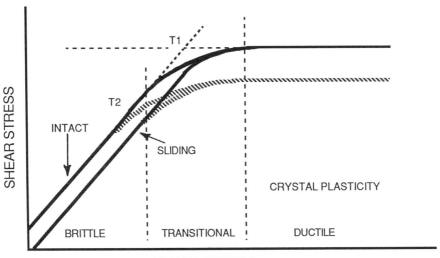

b.

Fig. 11. Conceptualized Mohr diagrams for the brittle-ductile transition of intact and pre-fractured rocks. a. The situation where the transition is to ductile behavior where cataclastic flow is the mechanism. The envelope for sliding is shown as zone of behavior as the stress conditions vary along the sliding surface with displacement and geometry. b. The case for a transition to crystal plastic ductility. T_1 and T_2 show the decrease of shear stress with increasing temperature.

PROBING POROUS MEDIA WITH 1ST SOUND, 2ND SOUND, 4TH SOUND, AND 3RD SOUND

David Linton Johnson[1], *T. J. Plona*[1], *and H. Kojima*[1,2]

[1]Schlumberger-Doll Research
Old Quarry Road
Ridgefield, CT 06877-4108

[2]Department of Physics
Rutgers, the State University
New Brunswick, NJ 08903

ABSTRACT

Historically, the understanding of superfluid ^{4}He in terms of the macroscopic two-fluid equations of motion was established in no small measure by means of experiments in porous media and other restricted geometries. It is now possible to turn this situation around and use the superfluid as a probe of the transport properties of porous media. We shall discuss specifically what one can learn from experiments using 1st, 2nd, 4th, and 3rd sound; we also explore the transferability of these results to other transport experiments. Many of the relevant geometrical parameters are those which arise in a canonical electrical conductivity problem in which the porous solid is insulating, the pore fluid is conducting, and there is an additional surface conductivity lining the walls of the pore space. The most important geometrical parameters are the three-dimensional tortuosity of the pore space, α_∞, the two-dimensional tortuosity of the pore/grain interface, α_2, and Λ, which is a well-defined measure of connected pore sizes.

I. INTRODUCTION

"Helium liquifies at 4.2°K. Between this temperature and 2.17°K this liquid (He I) behaves like an ordinary fluid. But as the temperature is lowered below $T_\lambda = 2.17°K$ it transforms into a second fluid phase (He II) which exhibits strikingly different dynamical properties[1]." It has been 45 years since Landau proposed the macroscopic two-fluid equations of hydrodynamics for He II; these equations have been oustandingly successful in understanding the macroscopic properties under a very wide range of conditions. Many of the relevant experiments were done in porous media and other restricted geometries. It is the purpose of the present article to utilize known properties of He II as a probe of porous media themselves.

The properties of He II are discussed extensively in the book by Putterman[1]. The acoustic properties are discussed in the review article by Rudnick[2]. It is *not*, however, necessary to understand these two references in order to understand the present article. Basically, and not to put too fine a point on it, He II behaves like a miscible mixture of two fluids: one is a normal fluid having a finite value of the viscosity and a density which is a function of temperature, $\rho_N(T)$; the other is the superfluid, having exactly zero viscosity, and a density $\rho_S(T)$. The total density $\rho = \rho_S + \rho_N$ is essentially independent of temperature but each component is highly temperature dependent; the experimentally determined[3] values of these two fractions is reproduced in Fig. 1.

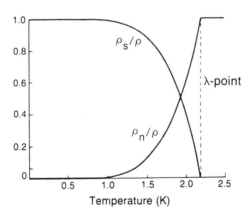

Figure 1. Experimentally determined superfluid and normal fluid fractions of He II below the λ point. Above T_λ there is no superfluid fraction and He I behaves like an ordinary fluid. From reference 3.

Because there are two degrees of freedom, the motion of the normal fluid and the motion of the superfluid, there are two distinct acoustic modes, called 1st and 2nd sound; the speeds of these waves as a function of temperature[4] are plotted in Fig. 2. 1st sound is an ordinary pressure/density wave whose speed is, to very high accuracy, given by the usual adiabatic expression:

$$c_1^2 = \frac{\partial P}{\partial \rho}\bigg|_s .\tag{1.1}$$

(In this article we shall utilize the fact that the thermal expansion coefficient is negligible.) 1st sound exists above and below the λ point, as one would expect. 2nd sound is a phenomenon which has no classical analogue. It is a temperature/entropy wave whose speed is given by

$$c_2^2 = \frac{\rho_S s_0^2}{\rho_N} \frac{\partial T}{\partial s}\bigg|_\rho ,\tag{1.2}$$

where s is the entropy per unit mass. Thus temperature and entropy variations are governed *not* by the heat equation but by the wave equation with a speed given by (1.2).

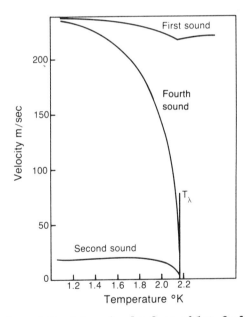

Figure 2. **Experimentally determined values of 1st, 2nd, and 4th sound as a function of temperature.** From reference 4.

Consider, now, a mode which exists whenever the normal fluid fraction is held immobile; this could occur in narrow capillary tubes in which viscous forces clamp the normal fluid. For historical reasons this has been termed 4th sound and its speed is given, to high accuracy, by

$$c_4^2 = \frac{\rho_S}{\rho} c_1^2 \quad . \tag{1.3}$$

This mode was experimentally discovered by Shapiro and Rudnick[5] who verified the prediction (1.3); its speed is also plotted in Fig. 2. In point of fact, 4th sound has always been done in porous media, not straight capillaries, and the experimentally observed speeds, v_4, are reduced from (1.3) by a geometrical factor, n, the "index of refraction": $v_4 = c_4/n$. Among other things we shall have a great deal to say about the significance of n.

There are several properties of He II which make it a powerful probe of acoustics in porous media and which we shall exploit in the present article:

1. It is several orders of magnitude more compressible than most solids, making the rigid frame approximation accurate to 1 part in 10^4, typically (Ref. 6).

2. The normal fluid fraction, ρ_N/ρ, can be made to be arbitrarily small simply by going to a low enough temperature; below 1.2°K, the normal fluid fraction is less than 1%. In effect, this means that the acoustics of He II can become identical with the acoustics of an ideal fluid.

3. Although the viscosity of the normal component, η, is not very temperature dependent, the effects of viscosity can be varied considerably by changing the temperature; because the normal fluid fraction decreases from unity to zero, the viscous skin depth, $\delta \equiv \sqrt{2\eta/\rho_N \omega}$, diverges at low temperature. [The viscous skin depth is the distance over which the vorticity decays to zero as one moves away from a bounding wall. See p. 88 of reference 7.] This means that one can probe porous media using δ as a yardstick whose size can be varied greatly as a function of temperature.

4. Since there are two distinct sound modes in He II one can extend the spectral range simply by switching from a 1st sound transducer to a 2nd sound transducer; in a typical resonant cavity experiment the wave length is fixed by the cavity dimensions, so that a change from 1st sound to 2nd changes the resonant frequency by a factor of ten (i.e. c_1/c_2).

5. Near a wall the superfluid density, ρ_S, rises from zero to its bulk value over a distance, $\xi_H(T)$, called the healing length[1]. The healing length diverges as one approaches the λ point from below. This means one can probe porous media using ξ_H as the yardstick; in practice, this applies to very small pores (less than 500Å).

6. He II can support a so-called 3rd sound wave that exists in thin films (≈ 100 Å). This means that one can probe the pore-solid interface with this mode.

The organization of this article is as follows: In Section II we make a detour and discuss a seemingly unrelated problem in electrical conductivity in porous media. The reason for this is that several of the key parameters for superfluid acoustics arise in that context. In Section III we discuss the response to an oscillatory pressure gradient of a rigid porous medium saturated with an ordinary Newtonian fluid. We utilize the results of these two sections to analyze the acoustic properties of porous media saturated with He II in Section IV; in this section it is assumed that macroscopic two-fluid hydrodynamics are valid on the scale of the pore sizes. The relevance of these results is, in turn, related to the acoustic properties of a compressible porous medium saturated with a compressible Newtonian fluid in Section V; we demonstrate the power of the Biot theory to predict the speeds and attenuations of all modes as a function of frequency with no adjustable parameters. We discuss healing length effects on 4th sound in porous media in Section VI. Section VII is devoted to a very brief description of 3rd sound in porous media. Our conclusions are summarized in Section VIII.

II. A CANONICAL PROBLEM IN ELECTRICAL CONDUCTIVITY

In this section we consider electrical conduction in a porous, insulating solid saturated with a conducting pore fluid, σ_f, and which also has an additional surface conductivity, Σ_s, which coats the pore walls. To do this we first assume that the pore fluid conductivity is some arbitrary function of position $\sigma(\bar{r})$. Suppose that a cube of this material, with edges of length L, is placed between bus bars across which there is a potential difference $\Delta\psi$. The microscopic potential $\psi(\bar{r})$ obeys the identity $\nabla \cdot [\psi(\bar{r}) \sigma(\bar{r}) \nabla \psi(\bar{r})] = \sigma(\bar{r}) |\nabla \psi(\bar{r})|^2$. Integrating over the volume of the pore space,[8] the effective conductivity is exactly related to $\sigma(\bar{r})$ and $\psi(\bar{r})$ by $\sigma_{\text{eff}} = |\Delta\psi|^{-2} L^{-1} \int \sigma(\bar{r}) |\nabla \psi(\bar{r})|^2 dV$. In a case of particular importance, the conductivity of the pore fluid is uniform [i.e. $\sigma(\bar{r}) \to \sigma_f$, $\psi(\bar{r}) \to \psi_0(\bar{r})$] and it is conventional to introduce the formation factor $F \equiv \sigma_f \sigma_{\text{eff}}^{-1}$; F is a geometrical parameter which is independent of σ_f:

$$\frac{1}{F} = \frac{\int |\nabla \psi_0(\bar{r})|^2 dV_p}{L |\Delta\psi|^2} , \quad (2.1a)$$

where the integration is over the pore space of the material. For our purposes it will be convenient to define the tortuosity of the pore space:

$$\alpha_\infty \equiv F\phi . \quad (2.1b)$$

[The reason for the subscript "∞" is consistency with a notation established earlier.[9]] It is a rigorous result that $\alpha_\infty \geq 1$. For straight channels of uniform cross-section $\alpha_\infty = \sec^2(\theta)$ where θ is the angle between the axis and the applied field. For an isotropic, homogeneous medium, $\alpha_\infty \geq (3-\phi)/2$, where ϕ is the porosity. For any cubic lattice (sc, fcc, bcc) of nodes connected with identically conducting straight tubes, $\alpha_\infty \equiv 3$.

Now, let the microscopic conductivity be perturbed to $\sigma(\bar{r}) = \sigma_f + \delta\sigma(\bar{r})$. There is a corresponding change, $\psi(\bar{r}) \to \psi_0(\bar{r}) + \delta\psi(\bar{r})$, in the microscopic potential, but, because $\delta\psi(\bar{r})$ vanishes on the bus bars, it can easily be shown that there is no first-order contribution to σ_{eff} from terms explicitly containing $\delta\psi$. [i.e. $\sigma_{\text{eff}} = F^{-1}\sigma_f + |\Delta\psi|^{-2} L^{-1} \int \delta\sigma(\bar{r}) |\nabla \psi_0|^2 dV + O[(\delta\sigma)^2]$. The first two terms are equivalent to the Born approximation in scattering theory.] Let us consider perturbations of the form $\delta\sigma(\bar{r}) = f(\epsilon)$ in which ϵ is a local coordinate measured from the pore wall into the conducting region. We assume the range of f is very small compared to the sizes of the pores, so that locally the walls appear to be flat on a length scale comparable to the range of f. $\Sigma_s \equiv \int f(\epsilon) d\epsilon$ is the interfacial conductivity and it follows that

$$\sigma_{\text{eff}} = F^{-1} \{\sigma_f + \frac{2\Sigma_s}{\Lambda}\} + O(\Sigma_s^2) , \quad (2.2)$$

where the quantity Λ has dimensions of length and is rigorously given by

$$\frac{2}{\Lambda} = \frac{\int |\nabla \psi_0(\vec{r}_w)|^2 \, dS}{\int |\nabla \psi_0(\vec{r})|^2 \, dV_p} \tag{2.3}$$

The integration in the numerator of (2.3) is taken over the walls of the pore-grain interface; that of the denominator is over the pore volume. Thus one accounts for the leading effects of surface conduction by means of a perturbation theory in which the relevant parameter, $2/\Lambda$, is an effective surface-to-pore-volume ratio wherein each area or volume element is weighted according to the local value of the field, $\vec{E}_0 = -\nabla \psi_0$, which would exist in the *absense* of the surface mechanism. This weighting *eliminates* contributions from isolated regions of the pore space that do not contribute to transport. Our motive for defining Λ in this manner is that in cases in which the pore space consists of non-intersecting, winding tubes of radius R, then $\Lambda \equiv R$.

We emphasize that Eqs. (2.2) and (2.3) are valid only if the thickness of the surface region is small compared to the pore size, *and if the effects of the surface conduction mechanism are small compared to that of the bulk*: $2\Sigma_s/\Lambda \ll \sigma_f$. In particular, (2.2) does not apply in the limit $\sigma_f \to 0$ because in this limit the surface mechanism, Σ_s, is the dominant one. It is straightforward to repeat the derivation given above, but with the assumption that σ_f is the perturbation on Σ_s. The result is

$$\sigma_{\text{eff}} = \frac{1}{f}[\Sigma_s + \frac{\lambda \sigma_f}{2}] + O(\sigma_f^2) \quad, \tag{2.4}$$

where the geometrical parameters f and λ are related to the microscopic field, $\vec{e}_0(\vec{r})$, which exists when $\sigma_f \equiv 0$,

$$\frac{1}{f} = \frac{\int |\vec{e}_0|^2 dS}{L |\Delta \psi|^2} \tag{2.5a}$$

$$\frac{2}{\lambda} = \frac{\int |\vec{e}_0(\vec{r}_w)|^2 \, dS}{\int |\vec{e}_0(\vec{r})|^2 \, dV_p} \quad . \tag{2.5b}$$

Note that $\vec{e}_0$ is well-defined in the pore space as the solution to Poisson's equation with Dirichlet boundary conditions. It shall be convenient to define the (dimensionless) tortuosity of the pore-wall surface by analogy with α_∞:

$$\alpha_2 \equiv f(S/V) \quad, \tag{2.5c}$$

where (S/V) is the surface-to-volume ratio. We have defined things so that the quantities S/V, f, α_2, and λ are in one-to-one correspondence with the quantities ϕ, F, α_∞, and Λ, respectively.

To illustrate these concepts, let us consider a specific class of models in which the porosity decreases by uniform growth of the insulating phase into the pore space. The growth of the grains can be viewed either as a change in F due to a change in ϕ *or*, equivalently, as a surface layer perturbation.[10] From (2.2) we have

$$\frac{2}{\Lambda} = -\frac{d \ln F}{d \ln \phi} \frac{S}{V_p} \equiv m(\phi) \frac{S}{V_p} , \qquad (2.6)$$

where ϕ is the porosity. Suppose, for example, we start in the high porosity limit with a simple cubic array of identical spheres. As the grain radius increases, the porosity decreases, and, once the grains overlap, the suspension evolves into the "grain consolidation" geometry.[11] At each value of the porosity we have numerically solved for F; by fitting a smooth curve through the calculated values of $F(\phi)$, it is straightforward to solve for Λ using (2.6). [We note that for the specific case of a random suspension of spherical grains it is experimentally known that $m(\phi) = 1.5 \pm 0.1$ for 40% $< \phi <$ 100%. Thus we conclude that

$$\Lambda = \frac{2\phi d}{9(1-\phi)} \qquad (2.7)$$

for suspensions of diameter d. See reference 10 for details.]

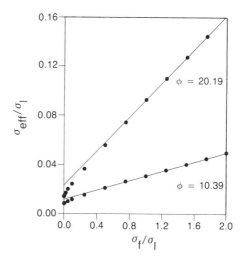

Figure 3. Calculated conductivity, σ_{eff}, as a function of pore fluid conductivity, σ_f for the grain consolidation model in arbitrary units. Two different grain radii, corresponding to two different porosities, are considered. The surface of the grains has been coated with a layer of conductivity σ_l and thickness $h = 0.015\, a$ (where a is the cube edge). The dots are the results of numerical calculations and the straight lines are based on Eqs. (2.2) and (2.6). From reference 10.

To investigate the crossover from (2.2) to (2.4) we performed numerical calculations on a grain consolidation model in which the grains were coated with a uniform layer of thickness h and conductivity σ_l, and the conductivity of the pore fluid, σ_f, was varied[10]: $\Sigma_s = (\sigma_l - \sigma_f)h$. Note that the numerical results plotted in Fig. 3 are in excellent agreement with the predictions of Eq. (2.2) (the straight line) for $\sigma_f \gtrsim \sigma_l$, but that they depart significantly for $\sigma_l \gg \sigma_f$ and ultimately hook onto (2.4). This departure is due entirely to the different *geometrical path lengths* associated with surface and bulk conductivity in porous media.

The most direct realization of this canonical problem occurs in the electrical conductivity of "shaly" sandstones in which the (insulating) grains are coated with appreciable amounts of clay minerals[12,13]. Typically, dry clay minerals contain charged impurities which are balanced by counter-ions bound to their external surfaces. However, once the pore space is saturated with brine, the hydrated counter-ions become mobile within a layer (whose thickness depends on the brine salinity) surrounding the clay particles. Surface conduction due to the counter-ions then proceeds in parallel with the ionic conductivity associated with the brine. For salinities greater than that of tap water the room-temperature counter-ion layer is less than 40 Å thick. As typical pore sizes are greater than 1000 Å, the first requirement for the validity of (2.2) is easily fulfilled. Empirically, one finds that for "large enough" pore-fluid conductivity, there is a linear relationship which is commonly written as[12,13]

$$\sigma_{eff} = F^{-1} \{\sigma_f + BQ_v\} \quad . \tag{2.8}$$

Here Q_v is the density of counter ions per unit pore volume and B is the equivalent conductance per ion, assumed to be the same as that of the bulk fluid (i.e. $\sigma_f = nB$, where n is the density of ions in the brine). Eqs. (2.2) and (2.8) are identical if $2/\Lambda$ is replaced by S/V_p because $Q_v = n_s(S/V_p)$ where n_s is the surface charge density of the clay mineral and $\Sigma_s = n_s B$. If the pore fluid conductivity, σ_f, is reduced below BQ_v then σ_{eff} departs from the linear prediction of (2.2) [Ref. 12]. It has been argued that this is due either to the breakdown of the assumption $h/\Lambda \ll 1$ (which seems unlikely), to a change in the surface ionic mobility (i.e. a change in B), or to the importance of the neglected diffusion current at low salinities[13]. Alternatively, we suggest that the decrease in σ_{eff} simply reflects the fact that (2.2) is valid only when $2\Sigma_s/\Lambda \ll \sigma_f$. The understanding of conductivity in shaly sands has been hampered by a lack of independent knowledge of the geometrical parameters $\alpha_\infty, \Lambda, \alpha_2, \lambda$, to say nothing of Σ_s and its dependence on pore-fluid salinity.

In the next sections we discuss how the parameters α_∞, α_2, and Λ relate to the acoustic properties of He II in porous media. In this regard we note that the properties of α_∞ were discussed at some length in the first Symposium[14].

III. DYNAMIC PERMEABILITY

One of the necessary ingredients for the understanding of superfluid acoustics in porous media is the dynamic permeability[9] i.e. the response of a simple Newtonian fluid to an oscillatory pressure gradient applied across a porous medium. To that end we consider a homogeneous, isotropic, porous solid which is perfectly rigid, either because it has a very large density, or very large elastic modulii, or both. This solid is saturated with an incompressible Newtonian fluid of density ρ_f and viscosity η. In practice, the results we shall obtain will apply to acoustics as long as the wave length of sound (in the fluid) is much larger than the characteristic sizes of pores and grains in the medium, ensuring that the fluid may be considered to be incompressible on the scale of the pore sizes. We assume that the properties of the fluid are unaffected by its proximity to the walls of the solid. We also assume the fluid has a negligible thermal expansion coefficient on any scale; this assumption ensures that pressure-density variations are decoupled from temperature variations (Ref. 7). In this section we shall consider some general properties of the dynamic permeability, defined below, and we shall use these properties to construct a simple, accurate, analytical model for the frequency dependence thereof. This model will depend on a few well-defined parameters which are functions of the geometry of the pore space.

The porous medium occupies the space $0 < x < L$, where L is extremely large compared to the sizes of the pores. The transverse (y,z) size of the system does not enter into the following, and may be taken as infinite, or as periodic, or as impermeable with finite extent. We apply a macroscopic pressure gradient, $\nabla P e^{-i\omega t}$, to the sample and follow the linear (i.e., small amplitude) response of the fluid to this applied gradient. The response is most conveniently defined in terms of the macroscopically averaged fluid velocity, $\bar{v}(\omega)$, which is defined so that $\phi \bar{v} \cdot \hat{n} A$ is the amount of fluid crossing a macroscopic surface of area A having an outward normal $\hat{n}$; since the area fraction actually in contact with fluid is ϕ, $\bar{v}$ represents a macroscopic fluid velocity. Under the stated assumptions $\bar{v}$ is obviously linearly related to the pressure gradient at any frequency:

$$\tilde{\alpha}(\omega) \rho_f \frac{\partial \bar{v}}{\partial t} = -\nabla P \quad , \tag{3.1a}$$

$$\phi \bar{v} = -\frac{\tilde{k}(\omega)}{\eta} \nabla P \quad . \tag{3.1b}$$

The frequency dependent tortuosity, $\tilde{\alpha}(\omega)$, is defined in (3.1a) by analogy with the response of an ideal (non-viscous) fluid for which $\tilde{\alpha}$ is real-valued and frequency independent and identical, in fact, with α_∞ defined earlier[15]. The frequency dependent permeability, $\tilde{k}(\omega)$, is defined in (3.1b) by analogy with the steady-state ($\omega = 0$) definition[16]. It has the dimensions of area (the conventional oil-field unit is the darcy: 1 darcy $= 10^{-8}$ cm^2). Obviously, the

$$\tilde{k}(\omega) = \frac{k_0}{\left\{1 - \frac{4i\alpha_\infty^2 k_0^2 \rho_f \omega}{\eta \Lambda^2 \phi^2}\right\}^{1/2} - \frac{i\alpha_\infty k_0 \rho_f \omega}{\eta \phi}} \quad (3.8b)$$

It is not readily apparent but (3.8) is the simplest such analytic function and it is difficult to think of others which satisfy the general requirements (1) through (4). It is straightforward to demonstrate that (3.8) agrees with (3.6) to better than 11% over the entire frequency range for real values of the frequency[9].

Let us, however, go one step further and consider the dynamic permeability of a genuinely disordered system formed from a simple cubic lattice of lattice constant l on which we randomly distribute tubes of radius r between the nodes thereof; the probability distribution is P(r). For cases in which P(r) is nontrivial, we have solved for the dynamic permeability by direct simulation on large (10x10x10) lattices and we have averaged over many (100) realizations as described elsewhere[22]; at any frequency, and for any individual tube, the fluid conductance is given by (3.5) in terms of the pressure difference between two connected nodes. The DC permeability, k_0, is determined by a separate simulation for $\omega = 0$; $\dot{V} = [\pi r^4/8\eta l] \Delta P$ for each tube. The porosity is simply $\phi = 3 \int \pi r^2 l P(r) dr/l^3$. As indicated in section II, α_∞ and Λ are determined from a separate simulation of the electrical conductivity problem in which the current in a given bond is related to the voltage difference between the nodes by

$$I = g(V_i - V_j) \quad , \quad (3.9a)$$

$$g(r) = \pi r^2/l \quad . \quad (3.9b)$$

Because of (2.1b), α_∞ is simply related to $\bar{g}$, the equivalent bond electrical conductivity of the random network,

$$\alpha_\infty = 3\frac{<g>}{\bar{g}} \quad , \quad (3.10a)$$

where $<g>$ is simply the arithmetic mean of the conductances

$$<g> = \int g(r) P(r) dr \quad . \quad (3.10b)$$

Equations (3.10a,b) follow directly from (2.1b) because $F = l/\bar{g}$. A very "tortuous" (i.e. disordered) network is one in which the effective conductance is much smaller than the mean, $\alpha_\infty \gg 3$. Having solved for is given by potential flow, $\vec{u}_p = -\nabla \psi_0$, for some ψ_0 which is identical, up to a multiplicative factor, to ψ_0 in Section II. Thus, the flow pattern is identical to that for an ideal fluid, except in the boundary layer which can be treated as a perturbation. The result is

$$\lim_{\omega \to \infty} \tilde{\alpha}(\omega) = \alpha_\infty \left[1 + \left[\frac{i\eta}{\rho_f \omega}\right]^{1/2} \frac{2}{\Lambda}\right] , \qquad (3.4a)$$

$$\lim_{\omega \to \infty} \tilde{k}(\omega) = \frac{i\eta\phi}{\alpha_\infty \rho_f \omega} \left[1 - \left[\frac{i\eta}{\rho_f \omega}\right]^{1/2} \frac{2}{\Lambda}\right] , \qquad (3.4b)$$

where α_∞ and Λ are *identical* with those quantities introduced earlier.

There exists a very simple example of an exactly solvable model which illustrates the general properties explicitly. The porous medium consists of fluid-saturated cylindrical tubes of radius R whose axes form an angle θ with the direction of the applied pressure gradient. The solution is given in different, but equivalent, forms by different authors[17-20]. We need the volume flow rate, $\dot{V}$, through a tube of length L having an imposed sinusoidally varying pressure drop, ΔP:

$$\dot{V} = \frac{\pi R^2}{i\omega \rho_f} \left[\frac{2J_1(KR)}{KR\, J_0(KR)} - 1\right] \frac{\Delta P}{L} \qquad (3.5)$$

where the J_i are Bessel functions. This expression has been verified experimentally in U-tube oscillations[19]. From (3.5) it is straightforward to deduce the dynamic permeability/tortuosity:

$$\tilde{\alpha}(\omega) = \left[\left[1 - \frac{2J_1(KR)}{KR\, J_0(KR)}\right] \cos^2\theta\right]^{-1} , \qquad (3.6a)$$

$$\tilde{k}(\omega) = \frac{\phi\eta}{i\omega\rho_f} \left[\frac{2J_1(KR)}{KR\, J_0(KR)} - 1\right] \cos^2\theta . \qquad (3.6b)$$

By inspection, $\tilde{\alpha}(\omega)$ and $\tilde{k}(\omega)$ have singularities only on the negative imaginary axis, and they obey the reflection symmetry (3.2); by considering the high and low frequency limits we can relate the general parameters to the specific:

$$\alpha_\infty = 1/\cos^2\theta , \qquad (3.7a)$$

$$k_0 = \phi R^2 \cos^2\theta/8 \, , \tag{3.7b}$$

$$\Lambda = R \, . \tag{3.7c}$$

(3.7a,b) are given in Scheidegger[16]; (3.7c) is obvious from the definition (2.3) because $\bar{u}_p(\bar{r}) = -\nabla \psi_0$ is constant in each tube.

In general, the parameters α_∞, k_0, and Λ are unrelated and are independently measurable, but if one can model the response of the system as a set of non-intersecting tubes canted at an effective angle θ, one may conjecture that the parameters are related to each other, at least approximately, by (3.7a-c):

$$M \equiv \frac{8\alpha_\infty k_0}{\phi \Lambda^2} = 1 \, . \tag{3.7d}$$

(If the porous medium can be modelled as canted slabs of fluid, for which Λ is equal to the width of the slab, then the 8 in (3.7d) is replaced by a 12.) A relationship equivalent to (3.7d), or modified by the introduction of dimensionless factors, is implicit in all previous approaches to this problem. For example, in the notation of Biot[18] the LHS of (3.7d) is equal to $\delta^2/8\xi$ where δ is the "structural factor" and ξ is the "sinuosity factor". In the notation of Attenborough[21], the LHS of (3.7d) is equal to n^2/s where n is the "dynamic shape factor" and s is the "static shape factor". If one makes reasonable guesstimates of the values of these dimensionless parameters, one is then able, in effect, to deduce values of Λ in terms of the other parameters ϕ, k_0, and α_∞. In the present article, we wish to emphasize that all parameters in (3.7d) are independently measureable and thus its approximate validity can be directly tested, which we shall do to a limited extent.

We return to the main theme of dynamic permeability. In order to analyse the acoustic properties of fluid-saturated porous media it is highly desirable to develop a model, in terms of simple analytic functions, for the frequency dependence of $\tilde{\alpha}(\omega)$. We require that the model satisfies the general properties (1)-(4) and that it depends only on the four parameters, α_∞, k_0, Λ, and ϕ which are well-defined for any porous medium; we do not assume that these four are related to each other by (3.7d). A very simple such form is

$$\tilde{\alpha}(\omega) = \alpha_\infty + \frac{i\eta\phi}{\omega k_0 \rho_f} \left\{ 1 - \frac{4i\alpha_\infty^2 k_0^2 \rho_f \omega}{\eta \Lambda^2 \phi^2} \right\}^{1/2} , \tag{3.8a}$$

two quantities are related to each other:

$$\tilde{\alpha}(\omega) = \frac{i\eta\phi}{\tilde{k}(\omega)\omega\rho_f} . \tag{3.1c}$$

The relationship between $\tilde{\alpha}(\omega)$ and $\tilde{k}(\omega)$ is roughly analogous to that between the dynamic electrical conductivity $\tilde{\sigma}(\omega)$ and the dynamic dielectric function $\tilde{\epsilon}(\omega)$ which arise in the electrodynamics of continuous media: $\tilde{\epsilon}(\omega) = 1 + 4\pi i \tilde{\sigma}(\omega)/\omega$.

Let us consider some of the properties of $\tilde{\alpha}/\tilde{k}$ considered as a function of ω in the complex plane. The following properties have been derived in some detail in reference 9 and will simply be quoted here.

1. Because the response of the system to a real-valued stimulus must itself be real-valued, we have the reflection symmetry:

$$\tilde{\alpha}(-\omega^*) = \tilde{\alpha}^*(\omega) , \tag{3.2a}$$

$$\tilde{k}(-\omega^*) = \tilde{k}^*(\omega) , \tag{3.2b}$$

where the asterisk signifies complex conjugation.

2. All of the singularities in $\tilde{\alpha}(\omega)$ or $\tilde{k}(\omega)$ occur on the negative imaginary axis: $\omega_s = -i|\omega_s|$.

3. Obviously, for low enough frequencies the dynamic permeability approaches its DC value:

$$\lim_{\omega \to 0} \tilde{k}(\omega) = k_0 , \tag{3.3a}$$

$$\lim_{\omega \to 0} \tilde{\alpha}(\omega) = \frac{i\eta\phi}{k_0 \rho_f \omega} , \tag{3.3b}$$

where k_0 is the real-valued permeability conventionally measured in an experiment in which the sample is subjected to a static pressure drop.

4. In the limit of high frequencies, the viscous skin depth, $\delta = \sqrt{2\eta/\rho_f\omega}$, eventually becomes much smaller than any characteristic pore size. From the microscopic Stokes equation the vorticity, $\nabla \times \vec{u}$, obeys the diffusion equation with a diffusion length given by δ. This means that any vorticity generated at the pore walls decays to zero as one moves away from the wall into the bulk of the pore (Ref. 7, p. 91). Therefore, except for a boundary layer of thickness δ, the fluid motion

the voltages, V_i, at each node we can very simply evaluate Λ from (2.3):

$$\Lambda = \frac{\sum\limits_{i,j}(V_i - V_j)^2 r_{ij}^2}{\sum\limits_{i,j}(V_i - V_j)^2 r_{ij}} \quad . \tag{3.11}$$

What sort of distribution, $P(r)$, ought we to consider? Clearly we wish one for which α_∞ is appreciably larger than 3. In a previous simulation of the electrical conductivity using a rectangular distribution of conductances

$$Q(g) = \begin{cases} \dfrac{1}{4999} & 1 < g < 5000 \\ 0 & \text{otherwise} \end{cases} \tag{3.12}$$

it was found that $\bar{g} = 2182$ for the simple cubic lattice[22]; this gives a value for the tortuosity, $\alpha_\infty = 3.44$, which is not very different from that for an ordered lattice. In order to increase the disorder we wish to use a distribution which emphasizes the smaller conductors; to be specific, we consider the distribution

$$P(r) = \frac{1}{c_0} e^{-r/c_0} \quad . \tag{3.13}$$

The porosity of this network is $\phi = 6\pi (c_0/l)^2$. We find the DC permeability to be $k_0 = (0.41 \pm 0.05)(c_0^2/l)^2$. (The error bars are statistical.) From the simulation of the electrical conductivity we find $\alpha_\infty = 7.94 \pm 0.47$; this value is considerably larger than those actually measured on a collection of fused glass bead samples[23,24] for which $10\% < \phi < 35\%$, giving us confidence that this simulation represents an appreciably disordered system. From the same electrical simulation we find $\Lambda = (0.93 \pm 0.10)c_0$.

Before we compare these numerical results against the model, (3.8), we should like to make two observations about the values of the parameters themselves: (a) The value of Λ is considerably smaller than that deduced from the specific area; we find $2/\Lambda = 2.15/c_0$ whereas the surface-to-pore-volume ratio is $1/c_0$. Evidently the weighting procedure implied by (2.3) substantially favors the smaller tubes. (b) As regards the approximate validity of (3.7d) with the values determined above, we find $8\alpha_\infty k_0/\phi\Lambda^2 = 1.6$ instead of 1.0. One can judge for oneself whether the glass is half full or half empty.

[As a further check of (3.7d) we have considered other distributions which also tend to emphasize the smaller conductors. An example is the "shrinking tubes" model[24] in which the pore space is pictured as a simple cubic array of tubes whose initial radii, $\{r\}$, are specified in terms of a distribution $P(r)$, and the porosity is then reduced by randomly shrinking tubes by a factor x. In Fig. 4 results are shown for the specific choices $P(r) = c_0^{-1} \exp(-r/c_0)$ and $x = 0.5$. Note that Λ decreases more rapidly as a function of porosity than the pore volume to surface ratio. Note also that, over a range of porosity in which k_0 varies by eight decades, the variation in M is about a factor of 2. We conclude that Λ, which derives from the solution to Poisson's equation, is closely linked to k_0, which derives from the solution to Poiseuille flow. Clearly, though, there must be a geometry for which this analogy breaks down.]

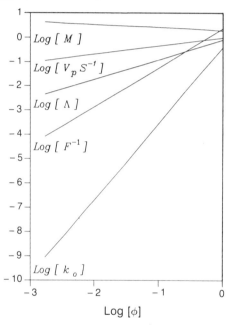

Figure 4. Variations in $M \equiv 8\alpha_\infty k_0/(\phi \Lambda^2)$, V_p/S [the pore-volume-to-surface area ratio], F, Λ, and k_o for the shrinking tubes model cross-plotted against the porosity. Λ and V_p/S are normalized to c_o, F is the formation factor times $c_o^2 l^{-2}$ (where l is the tube length) and k_o is the permeability multiplied by $l^2 c_o^{-4}$. Note that there is a strong correlation between Λ and k_0 via equation (3.7d) even as k_0 varies by eight orders of magnitude. From reference 10.

In Figs. 5a,b we plot the magnitude and phase, respectively, of the dynamic permeability deduced from the numerical simulations using the exponential distribution. We also compare these results with the predictions of the simple model, (3.8). Once again, we see that the model dynamic permeability agrees very well with that calculated directly; there are no adjustable parameters in the model, all of them having been calculated independently.

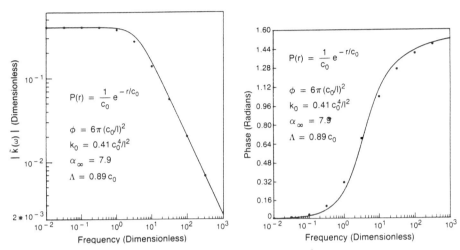

Figure 5. (a) Absolute value, in units of c_0^4/l^2, and (b) phase of the dynamic permeability of a random network of tubes distributed on the bonds of a simple cubic lattice using the exponential probability distribution of tube radii. The dimensionless frequency is $\omega \rho c_0^2/\eta$. The points are the results of direct simulations. The solid curves are the predictions of the model dynamic permeability, (3.8). The input parameters for the model, k_0, α_∞, and Λ, were determined by means of separate simulations, as explained in the text. From reference 9.

We hasten to point out that we have invented a model $\tilde{k}(\omega)$ which correctly matches the frequency dependence of the first two leading terms of the exact result for high frequencies, (3.4a,b), but only one term for low frequencies, (3.3a,b). Thus if we expand about $\omega \sim 0$ ve

$$\tilde{k}(\omega) = k_0 + i\omega\beta + O(\omega^2) \quad . \tag{3.14}$$

The point is that our model, (3.8), may very well yield values of β which differ substantially from the exact values. We refer the reader to a recent article by Norris[25] who has pointed out some rigorous theoretical statements that can be made about this coefficient as well as values thereof for some simply shaped pores. However, β appears not to be directly measureable in any experimental situation, except as a correction to the DC behavior. To some extent this has happened in the experiments considered by

Attenborough[21]; it would be interesting to compare the low frequency behavior of these systems against the predictions of our model, (3.8), using values of Λ deduced from superfluid ^{4}He experiments (Secs. IV and VI, below).

The conclusion of this section is that we have found a simple analytic expression for the frequency dependence of the permeability/tortuosity in terms of the well-defined (and experimentally measurable) high and low frequency parameters. The model is the rough equivalent of the Debye model for relaxation phenomena in dielectric properties. As compared against direct simulations on large networks, the model works well for a narrow distribution of tube radii, $P(r) = \delta(r-c_0)$, a wide distribution of tube radii, $P(r) = c_0^{-1} \exp(-r/c_0)$, and the bimodal distribution corresponding to percolation theory, $P(r) = p\, \delta(r-c) + (1-p)\delta(r)$. We conclude that unless the distribution is pathological (having a divergence at $r = 0$) the model will continue to give an accurate description of the frequency dependence of the permeability. Quite likely, other simple forms will work as well as the one considered here.

We also conclude, based on direct calculations of the parameters themselves, that the Λ parameter is closely related to the DC permeability, k_0, through (3.7d), at least for random lattice networks; the relationship (3.7d) may be violated by a factor of 2 but not, apparently, a factor of 10. This also seems to hold true for the available experimental data, which we discuss in Sections IV and VI.

IV. BULK SUPERFLUID ACOUSTIC MODES IN POROUS MEDIA

A. Theory

As per our introductory remarks, we may sensibly regard most porous solids as being rigid so that the acoustics is governed strictly by the motion of the ^{4}He. We assume the pores are large enough that the Helium motion is governed by the bulk equations of two-fluid hydrodynamics. The dominant mechanism for attenuation is the shear viscosity, η, of the normal fluid. If we look for plane-wave solutions, varying as $e^{i(\vec{q}\cdot\vec{r}-\omega t)}$, then it may be shown[9,26] that the dispersion relations, $q(\omega)$, are, to high accuracy,

$$q_-^2 = \frac{\omega^2}{c_1^2} \left[\frac{\tilde{\alpha}(\omega)\alpha_\infty}{\alpha_\infty + (\tilde{\alpha}(\omega) - \alpha_\infty)\frac{\rho_S}{\rho}} \right] . \qquad (4.1)$$

$$q_+^2 = \frac{\omega^2}{c_2^2} \left[\alpha_\infty + (\tilde{\alpha}(\omega) - \alpha_\infty)\frac{\rho_S}{\rho} \right] . \qquad (4.2)$$

$\tilde{\alpha}(\omega)$ is that appropriate for a hypothetical Newtonian fluid of density $\rho_N(T)$ and viscosity η. It is clear that measurements of the speed and attenuation of these modes enable one to directly deduce $\tilde{\alpha}(\omega)$ although the existing data are either for the strictly high frequency or strictly low frequency limits. Let us consider these two limits implied by (4.1,2).

<u>High Frequencies:</u> At sufficiently high frequencies (3.4) applies and so the dispersion in the phase velocity and the attenuation implied by (4.1) are

$$\lim_{\omega\to\infty} v_- = \frac{c_1}{\sqrt{\alpha_\infty}} \left[1 - \frac{\rho_N}{\rho} \frac{\delta(\omega,T)}{2\Lambda} \right] , \qquad (4.3a)$$

$$\lim_{\omega\to\infty} \frac{1}{Q_-} = \frac{\rho_N}{\rho} \frac{\delta(\omega,T)}{\Lambda} , \qquad (4.3b)$$

where $\delta = \sqrt{2\eta/\rho_N\omega}$ is the viscous skin depth in the normal fluid. Similarly, (4.2) gives

$$\lim_{\omega\to\infty} v_+ = \frac{c_2}{\sqrt{\alpha_\infty}} \left[1 - \frac{\rho_S}{\rho} \frac{\delta(\omega,T)}{2\Lambda} \right] , \qquad (4.4a)$$

$$\lim_{\omega \to \infty} \frac{1}{Q_+} = \frac{\rho_S}{\rho} \frac{\delta(\omega,T)}{\Lambda} \quad . \tag{4.4b}$$

In this limit the modes are essentially 1st and 2nd sound but with speeds reduced by the factor $\sqrt{\alpha_\infty}$ because of the tortuous, winding pore space. The temperature and frequency dependences in (4.3a,b) and (4.4a,b) have been reported by Singer, et al. (ref. 27) for five samples of fused glass beads. Baker[26,28] has also made similar measurements on a set of three samples of sintered bronze spheres. In particular, values of the parameter Λ ("r" in the notation of Singer, et al., and "$(8/\delta)\sqrt{k\alpha/P}$" in Baker's notation) have been measured. These are listed in Table I and compared against the theoretical values of Λ in suspensions, (2.7), which works surprisingly well considering that there are obvious clustering effects.

TABLE 1: Measured and calculated values of Λ and deduced permeability ratio M (see text).

TECHNIQUE	ϕ	Grain Diameter	$\frac{2\phi d}{9(1-\phi)}$	Λ (Experimental)	k_o (μm)^2	$\frac{8Fk_o}{\Lambda^2}$
1st Sound[a]	0.43	75 μm	12.6 μm	12.4 μm	7	1.36
1st Sound[a]	0.41	110 μm	17.0 μm	17.9 μm	12	1.44
1st and 2nd Sound[a]	0.41	500 μm	77.2 μm	53.2 μm	190	2.54
2nd Sound[b]	0.35	200 μm	24 μm	19 μm	---	---
4th Sound[c]	0.82	500 Å	506 Å	130 Å	---	---
4th Sound[c]	0.60	90 Å	30 Å	24 Å	---	---

[a] Ref. 28
[b] Ref. 27
[c] Ref. 40

Baker[28], in fact, has independently measured ϕ, α_∞, k_0 on his samples. He has experimentally deduced values for Λ from measurements of Q_-.

Does (3.7d) hold for those samples? In our notation the quantity $M \equiv 8\alpha_\infty k_0/(\phi \Lambda^2)$ of (3.7d) is $\delta^2/8$ in his. From Table 1 of Baker[28], $\delta^2/8$ varies from 1.4 to 2.5, which is comparable to the values which we found from the network simulations in Section III (i.e. 1.6 and 2.0). These values are also listed in Table I of the present article. This, then, is direct experimental evidence that the Λ parameter is closely related to DC permeability, k_0, in real porous media.

<u>Low Frequencies:</u> In this limit, the mode corresponding to 1st sound at high frequencies, (4.1), has a speed and attenuation given by

$$\lim_{\omega \to 0} V_- = \sqrt{\frac{\rho_S}{\rho}} \frac{c_1}{\sqrt{\alpha_\infty}} , \qquad (4.5a)$$

$$\lim_{\omega \to 0} \frac{1}{Q_-} = \frac{\alpha_\infty k_0 \rho_N^2 \omega}{\phi \eta \rho_S} . \qquad (4.5b)$$

This is 4th sound, a mode which exists whenever the normal component is clamped ($\bar{v}_N \equiv 0$) by virtue of its viscosity. It was first observed by Shapiro and Rudnick[5]. We note that the relationship between the index of refraction and the electrical tortuosity, $n = \sqrt{\alpha_\infty}$, was discussed[14] in the 1st Symposium. This crossover from 1st sound at high frequencies to 4th sound at low was predicted by Shapiro and Rudnick who did not, however, present a complete description of the effects of the porous medium i.e. their treatment is equivalent to the assumption that $\tilde{k}(\omega) = k_0$ is constant for all frequencies. Kriss[4] and, more recently, Tam and Ahlers[29], have experimentally verified the temperature and frequency dependence of (4.5b) though they did not report independent measurements of k_0 (which is expressed as $k_0/\phi = a^2/8$, in their notation).

In this low frequency limit, the 2nd sound mode becomes diffusive in character, $q_+^2 = i\omega/D$, with a diffusivity given by

$$D = \frac{\rho_N k_0}{\eta \phi} \left\{ \frac{\rho_S/\rho}{c_2^2} + \frac{\rho_N/\rho}{c_1^2} \right\}^{-1} . \qquad (4.6)$$

Like 2nd sound this is essentially a temperature/entropy wave. Theoretical descriptions of this mode have been presented by others using simplifying assumptions about the geometry of the porous medium[5,30,31]. This mode was unambiguously observed by Weichert and Passing[32] in plane-parallel capillaries of width 2d for which $k_0/\phi = d^2/3$. In a porous medium it ought to be straightforward to observe this mode using techniques similar to those of Chandler[33] but with 2nd sound transducers.

We stress that the parameters α_∞ and k_0/ϕ, which determine the speed and attenuation of 4th sound, as well as the diffusivity of the thermal wave, are easily measureable using standard techniques of electrical conductivity and fluid-flow resistance (Refs. 26, 28). In addition, α_∞ determines the high frequency speed of 1st sound[23] as well as that of 2nd sound[27].

B. Experiments

Following the leads of Baker and of Singer, et al., we have undertaken to probe a variety of porous media using techniques of superfluid acoustics described previously. A cylinder of each sample is carefully machined so that it fits snugly in a resonant cavity. The resonance frequencies are swept using either 1st or 2nd sound transducers, depending on whether one wishes to study the "+" mode (4.2), or the "−" mode, (4.1). The phase velocity, $V_p \equiv \omega/\text{Real}(q)$, of any mode is determined from the resonance frequencies in the usual manner: $L = \dfrac{jV_p(f_j)}{2f_j}$, where f_j is the resonance frequency of the j-th mode and L is the length of the sample. The Q of that mode is simply related to the width of the resonance peak, after the background signal has been removed. In Figures (6a,b) we show the index of refraction, $n_+ \equiv c_2 \text{Real}(q_+)/\omega$, and the specific attenuation, $\dfrac{1}{Q_+} \equiv \dfrac{2\,\text{Im}(q_+)}{\text{Real}(q_+)}$, of the second-sound mode in a commercially available porous medium; this sample was chosen because we were able to cover a wide range of temperature and frequency and were able to extend the measurements almost into the crossover region between the high and the low frequency limits implied by (3.3) and (3.4). The surface is that calculated using equation (4.2) and the model $\tilde{\alpha}(\omega)$ with the indicated values of the parameters; ϕ, k_0, and α_∞ were measured using standard laboratory techniques for porosity, fluid flow resistance, and electrical conductivity, respectively, whereas Λ is a free parameter. The dots represent actual data; a vertical line connects each datum with the theoretical value at the given frequency and temperature.

There are several points to be made from these figures. In the limit of low frequencies and/or low temperatures, where the mode becomes diffusive according to (4.6), one would expect the index, n_+, to diverge and the specific attenuation to level off at $1/Q_+ = 2$. The measured values of the phase velocities, which are reflected in n_+, are all in the high frequency limit where (4.4a) applies; indeed there is so little dispersion in the data that we have have simply verified $\lim_{\omega \to \infty} n_+ = \sqrt{\alpha_\infty}$ and we have little information on Λ, much less on the validity of the model, (3.8), throughout the crossover region. In this regard the data for the specific attenuation is more helpful; the largest value for $1/Q_+$ is 0.66 whereas the smallest is 0.01 thus providing fairly strong confidence in the deduced value of Λ. The values of the viscous skin depth $\delta = \sqrt{2\eta/\rho_N(T)\omega}$, range from 1.2 microns to 16 microns, the latter being quite comparable to Λ.

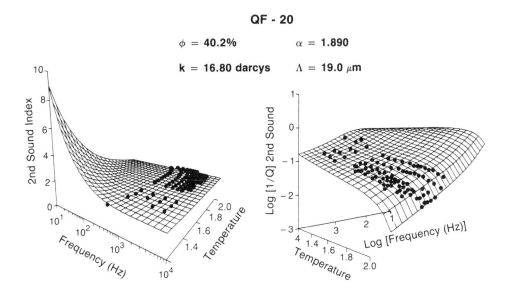

Figure 6. (a) Index of refraction, n_+, and (b) specific attenuation, $1/Q_+$, of the 2nd sound like mode in a porous medium manufactured by the Ferro company; QF-20 is a registered trademark. The surface is the theoretical prediction based on the model dynamic permeability, Eqn. (3.8) and the dispersion relation, Eqn. (4.2), using the indicated values of the four parameters; all but Λ were determined by independent measurements. There is a vertical line drawn from each datum to the surface.

It would be of great interest to be able to investigate the crossover region between the high and low frequency behaviors, (3.3) and (3.4) in order to probe, for example, the validity of the model (3.8) in that region. It is not possible to accomplish that goal with the 2nd sound resonances because those resonances become so broad that they overlap as the Q decreases toward 1/2. The "-" mode, which crosses over from 4th sound at low frequencies to 1st sound at high is a possible candidate because the Q's of those modes are always quite good. For the particular sample used in Fig. 6, however, all the 1st sound/4th sound resonances are approximately a factor of 10 *higher* in frequency (i.e. c_1/c_2) which is a step in the wrong direction; the only solution would be to shorten the sample length by a considerable amount. Alternatively, we have considered the crossover behavior in a sample of reduced pore size, a sample which is, in fact, a naturally occuring sandstone from the Berea formation. This sample has a DC permeability which is two orders of magnitude smaller than that of Figs. 6a,b, which is to say a pore size, Λ, one order of magnitude smaller. This in turn implies that the crossover frequency is much higher. The results for the index of refraction for the 1st-4th sound mode, $n_- = c_1 \text{Real}(q_-)/\omega$, as well as the specific attenuation, $1/Q_-$, are shown in Figs. (7a,b). We were able to observe only a few resonances at each temperature. Fortuitously, these resonances occur in the middle of the crossover region between 4th sound and 1st sound. There is excellent agreement between the theoretical predictions based on the model $\tilde{k}(\omega)$, Eq. (3.8), and the observed resonance frequencies. We hasten to add that the theoretical predictions in the crossover region are relatively insensitive to variations in Λ within the range 1-5μm; the other three parameters were measured independently, as before. The calculated values for the specific attenuation in the range $T < 1.6\,°K$ are all much smaller than those actually measured. This is quite possibly due to the fact that the mean free path for collisions becomes comparable to the pore sizes at low temperatures; as discussed by Baker[26] the use of the macroscopic viscosity becomes invalid. We note that at temperatures nearer the lambda point, where the macroscopic viscosity has meaning on the scale of the pore size, agreement between theory and experiment is quite good.

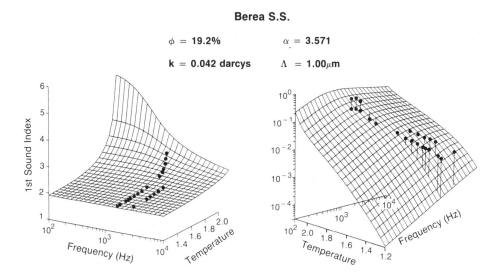

Figure 7. (a) Index of refraction, n_-, and (b) specific attenuation, $1/Q_-$, of the 1st-4th sound mode in a naturally occurring Berea sandstone. Conventions are as in Figs. 6a,b. The mode is 4th sound at low frequencies and 1st sound at high; the crossover frequency is highly temperature dependent because the viscous skin depth is temperature dependent at any frequency.

Our tentative conclusions so far are that the model response function, (3.8), together with the macroscopic equations of two-fluid hydrodynamics provides an excellent description of the acoustic modes in porous media saturated with He II. Specifically, we are able to extract values of Λ from the high frequency measurements, particularly from 2nd sound attenuation. We have seen an example of the breakdown of these ideas when the pores become comparable in size to the mean free path in the ^{4}He; we shall discuss another example of the breakdown of macroscopic two-fluid hydrodynamics in Section VI.

V. ACOUSTICS IN DEFORMABLE POROUS MEDIA

In this section we relax the assumption that the solid is not deformable and we consider the relevance of our results to the acoustic properties of porous media generally. It has become clear that the Biot theory[18,34] is the appropriate basic theory for such systems, although, of course, there are porous media that exhibit effects which are dominated by mechanisms outside the Biot theory; the equivalent statement for non-porous media is that standard elasticity theory, based on Hookian springs between atoms, is the appropriate basic theory of acoustics therein, although there are many mechanisms for attenuation and dispersion in real systems which lie outside standard elasticity proper. The basic idea of the Biot theory is that the average displacement of the fluid, $\vec{U}(\vec{r},t)$, and of the solid, $\vec{u}(\vec{r},t)$, are followed separately and on an equal footing, although the two motions are coupled. As a consequence, there are predicted to be two distinct longitudinal modes at all frequencies, a "Fast" wave and a "Slow" wave; although the former is propagatory at all frequencies, the latter is diffusive at low frequencies and propagatory at high. The theory has been successfully applied to such disparate systems as fourth sound in a superfluid/superleak system[6], pressure diffusion through porous media[33], slow waves and the consolidation transition[35], elastodynamics of gels[36], as well as the acoustic properties of "ordinary" porous media saturated with "ordinary" fluids[23]. Especially for these latter systems, the Biot theory has tremendous predictive power in the sense that a given sample can be characterized by direct and independent measurement of most of the input parameters and then used to predict the speeds and attenuations of the modes, regardless of what pore fluid is used. Aside from the densities and moduli of the fluid and solid constituents, the parameters are the porosity, ϕ, and the moduli of the skeleton frame, K_b and N, which can be deduced from the measured speeds of the dry material. The strictly low frequency properties are dependent on the value of the DC permeability, k_0, and the strictly high frequency properties are dependent on the values of the tortuosity, α_∞, and on the pore size, Λ. The distinction between high and low frequencies is, as before, whether the viscous skin depth is large or small compared to Λ. The predicted velocities have been verified experimentally in both limits. These and other properties of the theory are reviewed elsewhere[37,38].

The missing ingredient is a simple description of the viscous/inertial drag effects which can be expected to be valid over the entire frequency range in terms of a few independently measurable parameters characteristic of a given sample. In the Biot theory, these effects are described in terms of the relative motion between fluid and solid; the equation of motion for the fluid constituent can be written[23,28] as

$$\phi \rho_f \frac{\partial^2 \overline{U}}{\partial t^2} = \tilde{\rho}_{12}(\omega) \left[\frac{\partial^2 \overline{U}}{\partial t^2} - \frac{\partial^2 \overline{u}}{\partial t^2} \right] + \text{(spatial derivative terms)} \quad , (5.1a)$$

where

$$\tilde{\rho}_{12}(\omega) = -\left[\tilde{\alpha}(\omega) - 1\right]\phi \rho_f \quad . \quad (5.1b)$$

The quantity $\tilde{\alpha}(\omega)$ appearing in (5.1b) is independent of the elastic properties of the two constituents and is, in fact, identical to that defined in (3.1a). This can be seen by considering the Biot equations in the limit that the skeletal frame moduli, K_b and N, are much larger than the bulk modulus of the fluid, so that the solid does not move ($\overline{u} \equiv 0$); in this limit, (5.1a,b) reduce identically to (3.1a).

The implications are that the general properties of $\tilde{\alpha}(\omega)$ deduced in Section II automatically apply to the acoustics of porous media generally, via the Biot theory. Moreover, we have derived a simple theory, Eqn. (3.8), which gives a good description of $\tilde{\alpha}(\omega)$ in substantially disordered systems over the entire frequency range, as was shown in Figs. (5a,b) not to mention the experimental data in Figs. (6a,b) and (7a,b). One now has the confidence, once values for α_∞, Λ, k_0, and ϕ are measured, that there is a reliable means of calculating the acoustic properties over the full frequency spectrum.

For each sample our procedure, then, is to measure by independent means all of the parameters entering the Biot theory and use them to predict the speeds and attenuations of all three modes as a function of frequency; this can be done for any fluid one chooses for the pore fluid. The porosity, ϕ, frame modulii, K_b and N, the densities of the solid and fluid phases, and the modulus of the solid phase are determined as before. The tortuosity α_∞ is determined from the electrical conductivity, as before. The DC permeability is determined from conventional fluid-flow tests and Λ is determined from 2nd sound attenuation measurements. We then use the Biot theory an conjunction with the model dynamic permeability, $\tilde{k}(\omega)$, Eq. (3.8) to calculate the water-saturated speeds. A preliminary comparison of the theory against the experimental results, for a sample of "Ridgefield Sandstone" (fused glass beads of nominal diameter 200 μm) is presented in Fig. 8a,b. The experimental values were determined from immersion techniques described previously[23]. Experimentally we were unable to detect any dispersion in the phase velocities of any of the modes; from a theoretical point of view we do not expect any significant dispersion over the band width of our measurements. The observed attenuations of the fast and shear modes were so small that we were unable to quantify them; this is to be expected from the theoretical results. The observed attenuation of the slow wave is in excellent agreement with that predicted, although the band width of the measurements is only half a decade. [The observed increase at higher frequencies is due to the onset of Rayleigh scattering.] Thus, with the limited amount of data at our disposal, we have demonstrated the transferability of dynamic permeability deduced from acoustics in porous media saturated with He II to that in acoustics with a normal pore fluid (water).

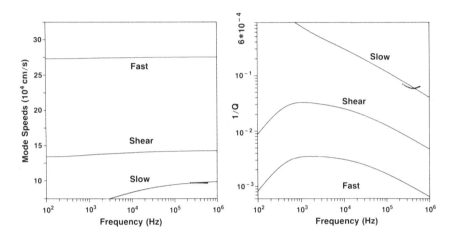

Figure 8. Acoustic properties of water-saturated Ridgefield Sandstone: (a) Phase velocity, and (b) specific attenuation, 1/Q, for the fast and slow compressional and the shear waves. The solid curves are theoretical predictions in which all the input parameters have been measured by independent means, as described in the text. The symbols represent measured values.

VI. HEALING LENGTH EFFECTS IN 4TH SOUND

In Section IV we discussed various modes in porous media saturated with He II under the assumption that the ^{4}He obeys the macroscopic two-fluid equations of hydrodynamics on the length scale of the individual pore. We saw that at high frequencies/large pores there are 1st and 2nd sound modes; in the limit of low frequencies/small pores the former crosses over to 4th sound. In this section we consider pores smaller still - so small, in fact, that there is significant departure from bulk behavior due to so-called healing length effects.

Assuming the validity of two-fluid hydrodynamics, let us write for the variation of the superfluid density near the walls: $\rho_S(\bar{r}) = \rho_S^{(o)}[1-g(\epsilon)]$, where $\rho_S^{(o)}$ is the bulk superfluid density and ϵ is the distance from the wall. The definition of the healing length is $\xi_H \equiv \int_0^\infty g(\epsilon)d\epsilon$. The temperature dependence of ξ_H has been deduced from experimental data on 3rd sound measurements[39] and is plotted in Fig. 9.

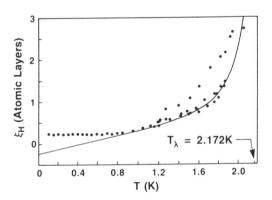

Figure 9. Experimentally deduced values of the healing length, ξ_H, as a function of temperature. One atomic layer = 3.6 Å. From reference 39.

This hydrodynamic problem also maps directly onto the canonical electrical problem and, to first order in $\dfrac{\xi_H}{\Lambda}$, the speed of 4th sound is

$$v_4^2 = \frac{1 - \dfrac{2\xi_H}{\Lambda}}{F\phi} \frac{\rho_S^{(o)}}{\rho} c_1^2 (1+\zeta) \quad , \tag{6.1}$$

where c_1 is the speed of 1st sound (assumed constant in the healing length region) and ζ is a parameter characteristic of bulk He II whose value is less than 0.01 for all temperatures at SVP. This result is equivalent to an equation used by Tam and Ahlers[40] to interpret their 4th sound data in packed powder superleaks using ξ_H calculated from a theory using neutron-scattering and thermodynamic data. Values of Λ ($2a/f_1$ in their notation) for two superleaks are given in Table I and compared against the theoretical values (2.7). While the geometry of packed-powder superleaks does not correspond to a random distribution of particles because of clustering effects, the calculated values are nevertheless close to those measured.

It is somewhat ironic that Λ appears in the result for the acoustic properties of systems with very large pores (viscous skin depth effects on 1st and 2nd sound) as well as for systems with very small ones (healing length effects on 4th sound).

VII. 3RD SOUND IN POROUS MEDIA

In the previous section we referred briefly to 3rd sound; this is a mode which exists in a film of ^{4}He which is thin enough that the normal fluid is clamped ($\delta \gg d$) but thick enough that the bulk equations of two fluid hydrodynamics still hold. If the van der Waals attraction between the ^{4}He atoms and the surface is the dominant restoring force then the speed of 3rd sound is simply

$$c_3 = \sqrt{\frac{\rho_S}{\rho} \frac{\alpha}{d^3}}, \qquad (7.1)$$

where α is the so-called Hamaker constant (*not* the tortuosity) and d is the film thickness. At low coverages there are depatures from (7.1) due to a variety of effects including a finite value of the healing length; at high coverages retardation effects set in. Putterman[1] gives an excellent description of many of these effects.

If one adsorbs Helium onto the internal surface of a porous medium and if one makes the assumption that the coverage is reasonably uniform, then it is straightforward to show that the speed of 3rd sound in the porous medium is simply

$$v_3 = \frac{c_3}{\sqrt{\alpha_2}}, \qquad (7.2)$$

where α_2 was defined in Section II in connection with the electrical conductivity problem. This result holds regardless whether c_3 is given by (7.1) or not; it is understood that the surface of the pore-wall is assumed to be flat locally and that the ^{4}He coverage is uniform so that c_3 is everywhere the same.

Rudnick and co-workers have performed a series of experiments on porous media partially saturated with He II. They used porous media for which the pore sizes are small compared to the viscous skin depth, so that the normal fluid is clamped, but large compared to the healing length. Some of their data for the index of refraction of the superfluid-based sound in packed powder superleaks is plotted in Fig. 10; this figure comes from an article[41] appearing in the 1st Symposium on the Physics and Chemistry of Porous Media. At an arbitrary filling fraction the sound mode is a complicated mix of 3rd sound, surface tension sound, and 5th sound; the manner in which the index, n, is extracted from the data is also somewhat complicated but is well described in Ref. 41; the value of the index is obviously dependent on the assumed value of the "bare" speed against which the experimenatal data is normalized. In the present article we simply point out that in the limit of low coverage the sound is essentially a surface mode which propagates in a film coating the walls of the solid; (7.2) applies and the index is obviously $\lim_{f \to 0} n = \sqrt{\alpha_2}$. In the limit of high coverage the mode is a bulk mode and one has $\lim_{f \to 1} n = \sqrt{\alpha_\infty}$. [In point of fact, for $f \neq 1$ the mode is 5th sound, not 4th, but the limiting value of the index is still $\sqrt{\alpha_\infty}$. See Ref. 41.] On an intuitive level it is obvious that $\alpha_2 > \alpha_\infty$ because a surface mode has to follow the convolutions of the surface but a bulk mode can "shoot across" each pore.

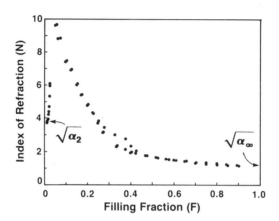

Figure 10. The index of refraction n of superfluid sound in a partially saturated porous medium under clamped conditions, as a function of the filling fraction [Ref. 41]. The method by which n is extracted from the data is described in this reference and references therein. Note that the index for low coverage is simply related to α_2 whereas that for high coverage is simply related to α_∞.

For completeness we simply mention that the physics of these sound waves for extremely low coverages, $f << 1$, is controversial at present. The controversy is between those who claim that a very dilute system behaves like a 3-dimensional Bose condensate and those who maintain that there is a 2-dimensional Kosterlitz-Thouless transition, modified by the local curvature of the interface. The data from which Fig. 10 was deduced involved coverages for which the film coverage was thick enough that these issues do not arise.

VIII. CONCLUSIONS

We should like to emphasize the following points:

1). We have developed some understanding of the general properties of the dynamic response function, $\tilde{\alpha}(\omega)/\tilde{k}(\omega)$ and have used these properties to construct a simple and successful "Debye-like" theory for the response over the entire frequency range. The model is specified in terms of 4 well-defined parameters which are characteristic of the geometry of the pore space and which are relevant to a variety of different exerimental contexts: ϕ, k_0, α_∞, and Λ. These last two are simply related to a canonical problem in electrical conductivity.

2). We have demonstrated how to measure $\tilde{\alpha}(\omega)$ using superfluid acoustics.

3). We have demonstrated how to use $\tilde{\alpha}(\omega)$ as an input to the Biot theory of acoustics in deformable, fluid-saturated solids. One is now able to measure all of the parameters in the theory by independent means and predict the speed and atenuation of the fast compressional, slow compressional, and shear modes as a function of frequency with no adjustable parameters.

4.) We have focussed attenuation on a new, well-defined parameter, Λ, which is characteristic of *dynamically* connected pore sizes, is transferable to other experiments, and for which there is an explicit formula which allows one to construct a theory thereof. Λ seems to be the relevant pore size parameter which correlates strongly with DC permeability, k_0.

5). Three of the parameters which are accessible via superfluid acoustics - α_∞, α_2, and Λ - are necessary geometrical parameters for an understanding of electrical conduction in shaly sands. In these systems there is appreciable surface conduction due to the mobility of counter ions bound to the surface of the clay minerals. At present it is not known whether the surface conduction is independent of the salinity of the pore water. i.e. whether Σ_s changes when one changes σ_f. Superfluid acoustic measurements should certainly enable one to separate geometrical effects from dynamic ones and help to provide an answer to questions of this sort.

Acknowledgements

We are most grateful for extensive interactions with our colleagues: J. Banavar, R. Dashen, B. Halperin, J. Koplik, A. Libchaber, and L. M. Schwartz.

REFERENCES

1. S. J. Putterman, *Superfluid Hydrodynamics*, (North- Holland Elsevier, Amsterdam, 1974).

2. I. Rudnick, *New Directions in Physical Acoustics*, Proceedings of the Enrico Fermi Summer School, Course LXIII (Academic, NY, 1976), p. 112.

3. K. R. Atkins, *Liquid Helium* (Cambridge, London, 1959).

4. M. Kriss, Ph.D. thesis, Dept. of Physics, UCLA, 1969 (unpublished).

5. K. A. Shapiro and I. Rudnick, Phys. Rev. **137**, A1383 (1965).

6. D. L. Johnson, Appl. Phys. Lett. **37**, 1065 (1980); ibid. **38**, 827 (1980)(E).

7. L. D. Landau and E. M. Lifshitz, *Fluid Mechanics*, (Pergamon, NY, 1959).

8. This derivation is identical to that of (2.2) in D. J. Bergman, Phys. Rep. **43**, 377 (1978).

9. D. L. Johnson, J. Koplik, and R. Dashen, J. Fluid Mech. (to be published).

10. D. L. Johnson, J. Koplik, and L. M. Schwartz, Phys. Rev. Lett. **57**, 2564 (1986).

11. J. N. Roberts and L. M. Schwartz, Phys. Rev. B **31**, 5990 (1985).

12. M. H. Waxman and L. J. M. Smits, Soc. Pet. Eng. J.: Trans. AIME, **243**, 107 (1968).

13. W. C. Chew and P. N. Sen, J. Chem. Phys. **77**, 2042 (1982); **77**, 4683 (1982); C. Clavier, G. Coates, and J. Dumanoir, Soc. Pet. Eng. J. **24**, 153 (1984).

14. T. J. Plona and D. L. Johnson, in *Physics and Chemistry of Porous Media*; ed. by D. L. Johnson and P. N. Sen; (AIP, NY, 1984).

15. D. L. Johnson and P. N. Sen, Phys. Rev. B **24**, 2486 (1981).

16. A. E. Scheidegger *Physics of Flow Through Porous Media*, 3rd Ed. (U. Toronto Press, Toronto, 1974).

17. A. Bedford, R. D. Costley, and M. Stern, J. Acoust. Soc. Am. **76**, 1804 (1984).

18. M. A. Biot, J. Acoust. Soc. Am. **28**, 168 (1956); ibid **28**, 179 (1956).

19. D. A. P. Jayasinghe, M. Letelier, and H. J. Leutheusser, Int. J. Mech. Sci. **16**, 819 (1974).

20. C. Zwikker, and C. W. Kosten, *Sound Absorbing Materials* (Elsevier, Amsterdam, 1949).

21. K. Attenborough, J. Acoust. Soc. Am. **73**, 785 (1983).

22. J. Koplik, J. Phys. C **14**, 4821 (1981).

23. D. L. Johnson, T. J. Plona, C. Scala, F. Pasierb, and H. Kojima, Phys. Rev. Lett. **49**, 1840 (1982).

24. P.-z. Wong, J. Koplik, and J. P. Tomanic, Phys. Rev. B **30**, 6606 (1984).

25. A. Norris, J. Wave Mat. Int. (to be published).

26. S. R. Baker, Ph.D. thesis, Dept. of Physics, UCLA (1985) (unpublished).

27. D. Singer, F. Pasierb, and H. Kojima, Phys. Rev. B **30**, 2909 (1984).

28. S. R. Baker and I. Rudnick, IEEE Trans.-UFFC **33**, 118 (1986); Proc. 1985 IEEE Ultrasonics Symposium (to be published).

29. W. Y. Tam and G. Ahlers, J. Low Temp. Phys. **58**, 497 (1985).

30. G. L. Pollack and J. R. Pellam, Phys. Rev. **137**, A1676 (1965).

31. M. Weichert and L. Meinhold-Heerlein, J. Low Temp. Phys. **1**, 273 (1970).

32. M. Weichert and R. Passing, Phys. Rev. B **26**, 6114 (1982).

33. R. N. Chandler, J. Acoust. Soc. Am. **70**, 116 (1981).

34. M. A. Biot, J. Appl. Phys. **33**, 1482 (1962); J. Acoust. Soc. Am. **34**, 1254 (1962); M. A. Biot and D. G. Willis, J. Appl. Mech. **24**, 594 (1957).

35. D. L. Johnson and T. J. Plona, J. Acoust. Soc. **72**, 556 (1982).

36. D. L. Johnson, J. Chem. Phys. **77**, 1531 (1982).

37. R. D. Stoll, *Physics of Sound in Marine Sediments*, edited by L. Hampton (Plenum, NY, 1974).

38. D. L. Johnson, *Frontiers of Physical Acoustics*, Proceedings of the Enrico Fermi Summer School, Varenna, Italy, 1984 (North Holland-Elsevier, NY, 1986).

39. J. H. Scholtz, E. O. McLean, and I. Rudnick, Phys. Rev. Lett. **32**, 147 (1974).

40. W. Y. Tam and G. Ahlers, Phys. Lett. **92A**, 445 (1982).

41. S. Baker, J. Marcus, G. A. Williams, and I. Rudnick, in *Physics and Chemistry of Porous Media*, ed. by D. L. Johnson and P. N. Sen, (AIP, NY, 1984).

V. Electrical Properties

ELECTRICAL PROPERTIES FROM 10^{-3} TO 10^{+9} HZ -- PHYSICS AND CHEMISTRY

G. R. Olhoeft
U. S. Geological Survey, Denver, CO 80225

ABSTRACT

In dry materials, physical factors control all electrical properties. The addition of a polar liquid solvent such as water or alcohol adds a host of solvent-rock chemical interactions. These chemical interactions range from oxidation-reduction corrosion, cation exchange, and clay-organic processes at frequencies below 1 Hz to diffusion-limited relaxation around colloidal particles at frequencies up to 100 MHz. Most mixing formulas are based upon physical mixing of noninteracting materials, and they fail when chemical processes appear. If the specific chemical processes are identifiable, combined physical and chemical mixing formulas must be used. The simplest systems to model are noninteracting physical mixtures of solvents with pure silica sand. The most complicated systems are mixtures of solvents with chemically surface-reactive materials like clays and zeolites.

INTRODUCTION

The mechanisms of electrical charge transport (conduction) and of charge separation (polarization) vary considerably over frequencies of interest in geophysics. Deep magnetic sounding, magnetotelluric, and induced polarization measurements occur at frequencies below 1 Hz. A variety of electromagnetic induction techniques operate up to 1 MHz. Electromagnetic propagation techniques are used from 1 MHz to over 10 GHz. At frequencies above 100 MHz, physical mixing of noninteracting materials is adequately described by the Bruggeman-Hanai-Sen formula (1, 11). However, the BHS formula cannot yield dielectric permittivities that are higher than those of any pure end-member component of the mixture, nor can the BHS formula predict losses when scattering is present (when the scale of electromagnetic wavelength in the material approaches the scale of material inhomogeneity; 10).

Below 100 MHz, colloidal-sized particles with reactive surfaces in contact with polar solvents create new polarization phenomena that cannot be described or predicted by the BHS formula. The BHS formula fails because the components in the mixture interact with each other, yielding electrical properties greater than any end member component. This is most easily demonstrated with mixtures of sand, clay and water. Kutrubes (2) gives examples using sand and clay with benzene and methanol.

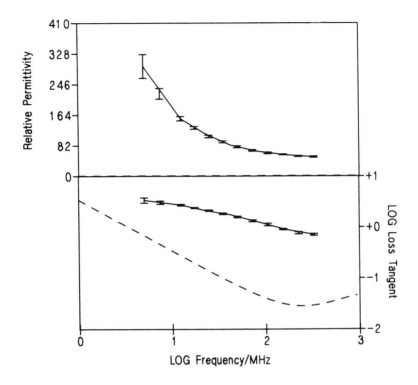

Figure 1. The frequency dependence of the dielectric properties of 22.9 weight percent clay mixed with 77.1 weight percent water at a bulk density of 1.049 g/cm^3. The sample key is in Table I. The solid line is the Hilbert test after Olhoeft (3, 4) and the dashed line is the lowest loss accurately measured by the sample holder and network analyzer (see text). The error bars are the data.

EXPERIMENTAL EVIDENCE

Figures 1 through 5 illustrate the high frequency dielectric properties of sand-clay-water mixtures (as described in Table I). The measurements were performed using GR-900 coaxial sample holders and an HP-8507 automatic network analyzer system (2). The dashed line in each figure represents the minimum loss that can be accurately measured by the system (limited by the residual losses in the sample holder and the repeatability of the network analyzer system). The error bars show the spread of the data from full two-port measurements. The solid lines in each figure are the expected values from the Hilbert Transform test of nonlinearity of the data (3, 4). Any significant difference between the plotted data and the solid line indicates Hilbert Distortion (3, 4, 6). At these frequencies and within experimental error, no Hilbert Distortion is observed.

Table I. Sample and Figure Key

Figure	DC Resistivity ohm-meters	100 MHz Relative Dielectric Permittivity & Loss Tangent		Bulk Density g/cm^3		Sand	Clay —weight percent—	Water
				low	high			
Figure 1 & 11	3.4	62.6	1.14	1.150	1.049	0.0	22.9	77.1
Figure 2 & 13	4.7	49.7	1.00	1.400	1.290	29.3	13.9	56.8
Figure 3 & 14	15.7	16.4	0.91	1.866	1.661	79.5	4.9	15.6
Figure 4 & 15	33.5	18.1	0.40	1.904	1.758	76.7	1.9	21.4
Figure 5 & 12	269.	12.1	0.088	2.087	1.633	83.6	0.0	16.4
Figure 8	4.4	52.8	1.00	–	1.084	17.9	16.5	66.4
Figure 8	6.6	32.3	1.09	–	1.636	58.2	10.1	31.7
Figure 8	13.4	20.2	0.84	–	1.808	77.5	4.0	18.5
Figure 8	125.	14.0	0.14	–	1.732	80.7	0.0	19.3

NOTES: Under bulk density, low and high refer to the density in the sample holder for the low (Figures 11-15) and high (Figures 1-10) frequency measurements. Sand is pure silica sand, clay is Clay Mineral Society CMS-SWy-1 Na-montmorillonite from Wyoming (15), and water is pharmaceutical-grade distilled, de-ionized water. The additional data used in Figures 7 through 10 are shown for Figure 8.

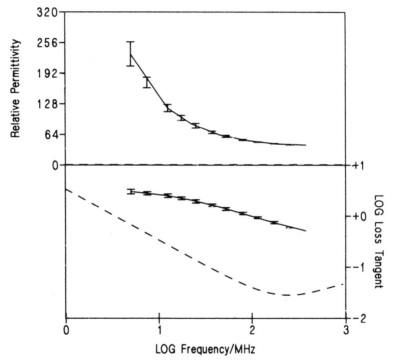

Figure 2. The dielectric properties of 29.3 weight percent sand, and 13.9 weight percent clay, and 56.8 weight percent water at a density of 1.290 g/cm^3.

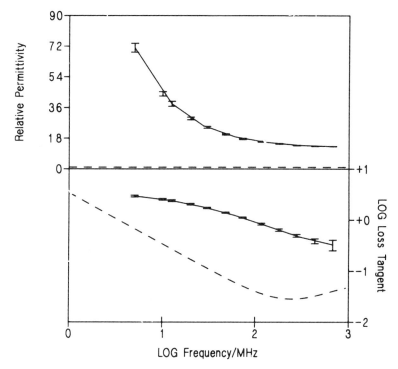

Figure 3. The dielectric properties of 79.5 weight percent sand, and 4.9 weight percent clay, and 15.6 weight percent water at a density of 1.661 g/cm^3.

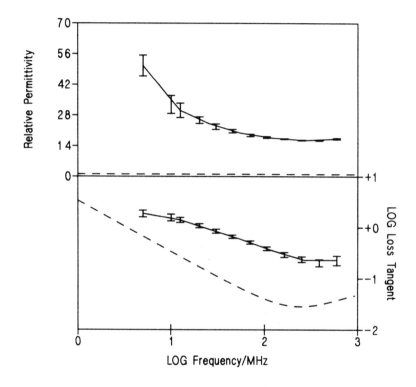

Figure 4. The dielectric properties of 76.7 weight percent sand, and 1.9 weight percent clay, and 21.4 weight percent water at a density of 1.758 g/cm^3.

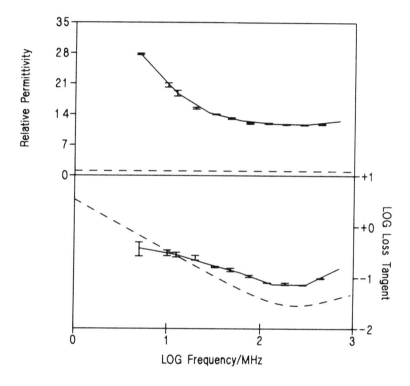

Figure 5. The dielectric properties of 83.6 weight percent sand, and 16.4 weight percent water at a density of 1.633 g/cm^3.

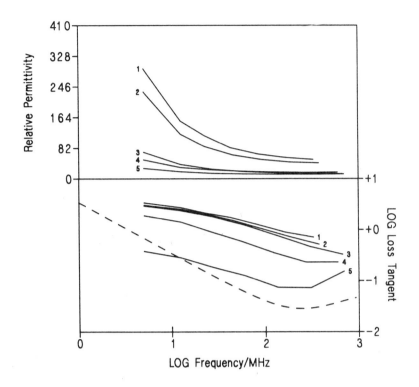

Figure 6. The data from Figures 1 through 5 overlain in one figure. The curve labels correspond to the figure numbers.

Figure 6 summarizes the data in Figures 1 through 5. Note the increasing loss tangent and increasing low-frequency dielectric permittivity with increasing clay content. Ground penetrating radar shows penetration of about 30 meters through clean, clay-free sand near 100 MHz (5, 7, 10). The addition of 5 weight percent clay increases the loss and decreases the penetration by a factor of 20.

Figures 7 through 10 illustrate the data from Figures 1 through 5 and additional data for other mixtures (Table I) plotted at fixed frequencies versus volume percent water and silica. In these figures, silica means total volume percent sand plus clay. The lines represent the prediction of the Hashin-Shtrickman (H-S) bounds for the Bruggeman-Hanai-Sen (BHS) formula with an exponent of 1/3 using relative dielectric permittivities of 4.5 and 80 for pure silica and water respectively. The solid lines are the BHS predictions for 300 ohm-m resistivity, and the dashed lines are for 3 ohm-m resistivity. The ellipses (some look like circles) are the measured data (with the vertical ellipse axis representing the data error), and the "error bars" are the H-S BHS predictions at the resistivity of each sample. At 300 MHz, the data fall within the predictions fairly well. The slightly high predictions in the high water content samples result from the high clay content that binds

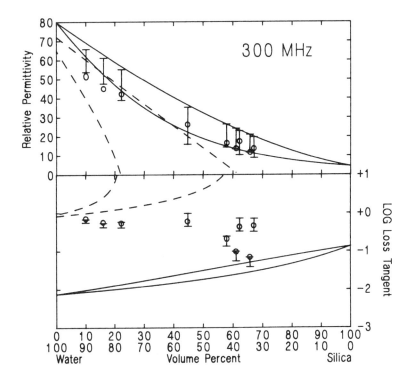

Figure 7. The data at 300 MHz from Figures 1 through 5 plus additional data at other sand-clay-water ratios along with predictions of the BHS formula (see text).

a significant fraction of the water. Thus, the plotted data should properly fall to the right (towards lower water content) to account for a more realistic amount of unbound water. This would move the data closer to the predicted values (14).

Diffusion-limited chemical reactions at the surfaces of particles have speeds (and hence frequency dependent relaxations) that are controlled by particle size (9, 16). At 100 MHz, the clay is beginning to have particle sizes which result in significant diffusion-limited chemical activity. This causes the BHS formula to under-predict the permittivity because it does not include the polarization from the chemical interaction between the materials (water and clay). At lower frequencies, the diffusion-limited processes become stronger, and the BHS formula is overwhelmed by chemistry.

Below 1 MHz, other electrical measurement techniques and the nomenclature of complex resistivity are commonly used (3, 6). Figure 11 illustrates the complex resistivity of saturated clay over the entire frequency range (a combination of the data in Figures 1 and 12). The closed circles are the resistivity and the

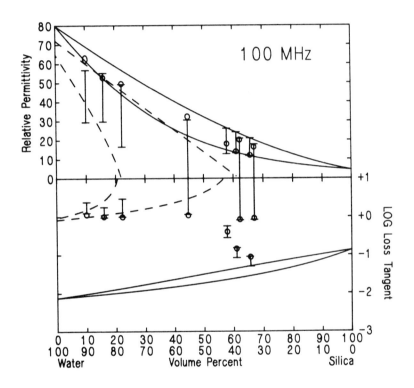

Figure 8. The data at 100 MHz as for Figure 7. Note that three of the H-S bounds bars hit the axis and do not cross from one part of the figure to the other.

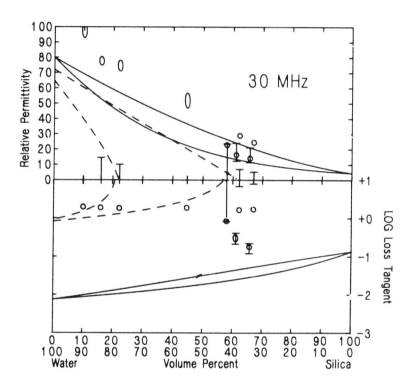

Figure 9. The data at 30 MHz as for Figure 7.

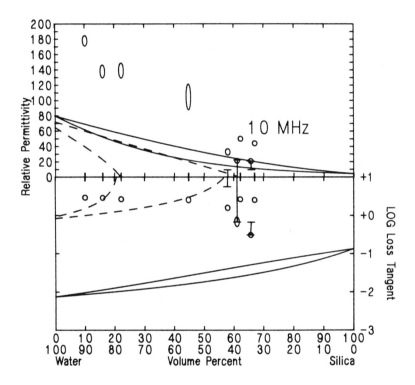

Figure 10. The data at 10 MHz as for Figure 7. Note that the two data points that fall within their H-S BHS bounds contain no clay, and the relative content of clay compared to sand increases with increasing water content.

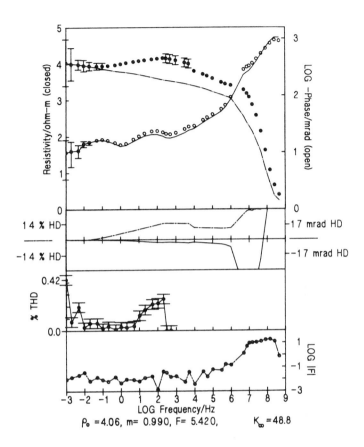

Figure 11. The complex resistivity spectra of 22.9 weight percent clay and 77.1 weight percent water (see text). This figure is a combination of Figures 1 and 12.

open circles are the phase angle. The dot-dash and solid lines are the predictions of the real and imaginary parts of the complex resistivity through the Hilbert transform (3, 4, 6). Deviations between the lines and the data are Hilbert Distortion (HD) and represent nonlinearity which indicates chemical reactivity. THD is total harmonic distortion, which is another indicator of nonlinearity and chemical reactivity. THD tends to indicate more oxidation-reduction reactions whereas HD indicates more cation exchange reactions (3, 4, 6, 7, 8). F is the Faradaic time constant distribution normalization factor and m is the volume chargeability (described in detail in Olhoeft; 6). At the bottom of the plot, ρ_o is the DC resistivity and K_∞ is the high frequency limiting relative dielectric permittivity.

Figure 12 compares the complex resistivity spectra of wet sand (circles) and wet clay (triangles), corresponding with the

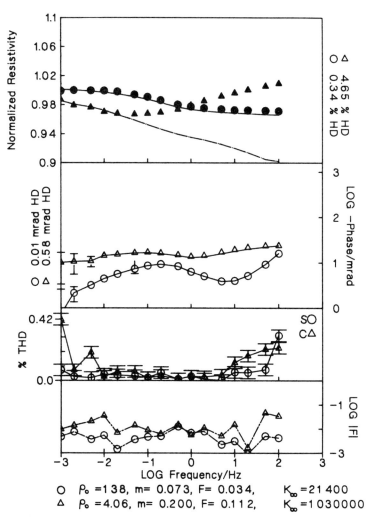

Figure 12. The complex resistivity spectra of wet sand (circles) and wet clay (triangles) as for Figures 1 and 5. Note the differences in resistivity and Hilbert Distortion (HD).

mixtures used in generating Figures 1 and 5 at higher frequencies. In this and succeeding figures, the resistivities are normalized to the DC resistivity of the sample (noted at the bottom of each figure). Note that the largest differences between the sand and clay are in the magnitude of the resistivity (4.06 in clay versus 138. in sand) and in the Hilbert Distortion (4.65 in clay versus 0.34 in sand). Figures 13 through 15 illustrate the complex resistivity of the intermediate sand-clay-water mixtures as for Figures 2 through 4. Note the abrupt decrease in resistivity and increase in Hilbert Distortion with a few weight percent clay.

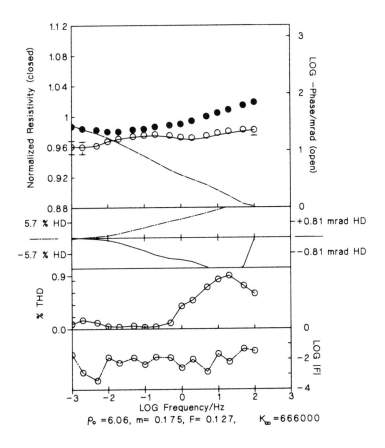

Figure 13. The complex resistivity spectra of 29.3 weight percent sand, 13.9 weight percent clay and 56.8 weight percent water at a density of 1.400 g/cm^3.

The montmorillonite clay has a surface area 2 orders of magnitude larger than the sand (15). The few weight percent clay completely coat the sand grains. As the dominant chemical reactions are surface processes, the clay coatings on the sand make the sample rapidly approach the properties of 100 percent clay.

DISCUSSION

In the high frequency data above 100 MHz, the sand-clay-water mixtures essentially appear as if there were no difference between the sand and clay (though there is a slight lessening in the apparent water content due to clay-bound water). Such mixtures are adequately modelled by noninteractive physical mixing such as the Bruggeman-Hanai-Sen formula (1, 11). Below 100 MHz, the highly reactive surface of the clay particles and their colloidal size range produce dielectric relaxation losses from diffusion-limited

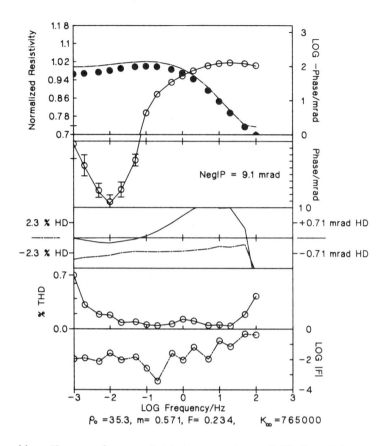

Figure 14. The complex resistivity spectra of 79.5 weight percent sand, 4.9 weight percent clay and 15.6 weight percent water at a density of 1.866 g/cm^3.

chemical reactions. Colloidal pyrite and gold are also known to cause high frequency relaxations (17). Such processes require more complicated mixing models that include the effects of chemical interactions between materials (1, 9). Below about 10 Hz, diffusion-limited chemical reactions around larger particles produce electrochemical relaxations commonly observed in the induced polarization method and for which more complicated models exist (16). Kinetics-limited processes such as cation exchange and clay-organic processes appear as electrical relaxations to frequencies below 0.001 Hz (6, 7, 8). These low frequency electrical properties may be more dominated by chemical than physical processes, requiring mixing formulas derived from electro-chemistry (12, 13). In the absence of chemically reactive materials, the physical formulas such as BHS work well at both high and low frequencies.

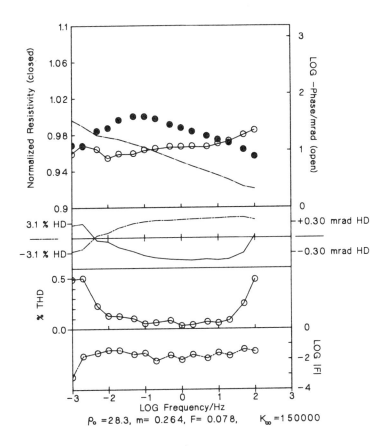

Figure 15. The complex resistivity spectra of 76.7 weight percent sand, 1.9 weight percent clay and 21.4 weight percent water at a density of 1.904 g/cm^3.

REFERENCES

1. Duhkin, S.S., Dielectric properties of disperse systems: in Surface and Colloid Science, v.3, E. Matijevic, ed., NY, Wiley, p.83-166 (1971).
2. Kutrubes, D.L., Dielectric permittivity measurements of soils saturated with hazardous fluids: M.Sc. Thesis, Dept. of Geophysics, Colo.School of Mines, Golden, 300p. (1986).
3. Olhoeft, G.R., Electrical properties: in Initial Report of the Petrophysics Laboratory, U.S.Geol.Survey Circular 789, p.1-25 (1979).

4. Olhoeft, G.R., Nonlinear electrical properties: in Nonlinear Behavior of Molecules, Atoms and Ions in Electric, Magnetic or Electromagnetic Fields, L. Neel, ed., Amsterdam, Elsevier, p.395-410 (1979).
5. Olhoeft, G.R., Applications and limitations of ground-penetrating radar: in Expanded Abstracts, Soc.Explor.Geophys. 54th Annual Int'l.Meeting and Expo., Atlanta, GA, p.147-148 (1984).
6. Olhoeft, G.R., Low-frequency electrical properties, Geophysics, v.50, p.2492-2503 (1985).
7. Olhoeft, G.R., Direct detection of hydrocarbon and organic chemicals with ground penetrating radar and complex resistivity: in Proc. Nat.Water Well Assoc. Conf. on Petroleum Hydrocarbons and Organic Chemicals in Ground Water, Nov.12-14, Houston (1986).
8. Olhoeft, G.R., Electrical conductivity: in Proc. of the 8th Workshop on Electromagnetic Induction in the Earth and Moon, Review Papers, Int.Assoc.Geomag. and Aeronomy, August, Neuchatel, Switzerland, p.2-1 to 2-13 (1986).
9. Pethig, R., Dielectric and Electronic Properties of Biological Materials, NY, Wiley, 376p. (1979).
10. Schaber, G.G., McCauley, J.F., Breed, C.S., and Olhoeft, G.R., Shuttle imaging radar -- physical controls on signal penetration and subsurface scattering in the Eastern Sahara: IEEE Trans. Geosci.& Rem.Sens., v.GE-46, p.603-623 (1986).
11. Sen, P.N., Scala, C., and Cohen, M.H., A self-similar model for sedimentary rocks with application to the dielectric constant of fused glass beads: Geophysics, v.46, p.781-795 (1981).
12. Sluyters-Rehbach, M. and Sluyters, J.H., Sine wave methods in the study of electrode processes: in Electroanalytical Chemistry, v.4, A.J.Bard, ed., NY, Marcell-Dekker, p.1-128 (1970).
13. Sluyters-Rehbach, M. and Sluyters, J.H., A.C. techniques: in Comprehensive Treatise of Electrochemistry, v.9, Electrodics-- experimental techniques, E.Yeager, J.O'M.Bockris, B.E.Conway, and S.Sarangapani, eds., NY, Plenum, p.177-292 (1984).
14. Ulaby, F.T., Moore, R.K., and Fung, A.K., 1986, Microwave remote sensing: active and passive, v.3, from theory to applications, Dedham, Artech House, p.1065-2162 (1986).
15. van Olphen, H. and Fripiat, J.J., eds., Data Handbook for Clay Minerals and Other Non-Metallic Minerals, NY, Pergamon, p.19-21 (1979).
16. Wong, J., An electrochemical model of the induced-polarization phenomenon in disseminated sulfide ores: Geophysics, v.44, p.1245-1265 (1979).
17. Young, G.N. and Johnston, D.H., The association of microfine pyrite with gold in Carlin-type deposits as determined by complex resistivity measurements: in Expanded Abstracts, Soc. Explor.Geophys. 56th Ann.Int'l. Meeting and Expo., Houston, p.167-168 (1986).

PERCOLATION IN OIL-CONTINUOUS MICROEMULSIONS

S. Bhattacharya, J. P. Stokes, M. J. Higgins,
M. W. Kim and J. S. Huang
Corporate Research Science Laboratories
Exxon Research and Engineering Company
Annandale, NJ 08801

ABSTRACT

Percolation in oil continuous microemulsions containing surfactant covered water droplets demonstrate behavior significantly different from standard percolation phenomena. The hopping of charge carriers (anionic surfactants) on the cluster of water droplets which rearrange in time results in a dynamic (or "stirred") percolation that yields critical exponents and frequency scaling different from static percolation in systems with quenched disorder.

Percolation transition in physical systems is a subject of considerable current interest. Anomalies in the electrical properties of insulator-metal or metal-superconductor composites and in mechanical properties in polymer gels have been studied extensively in recent years [1].
The observation of percolation in a microemulsion system [2] has attracted a great deal of attention because this system is significantly different. Microemulsions are self-assembling, interacting, equilibrium systems consisting of oil, water and surfactant. Percolation is observed in the globular phase of microemulsions consisting of surfactant-covered water droplets in an oil continuous medium. Charge transport is thought to occur via the hopping of anionic surfactants on a cluster of water droplets. In contrast with standard systems with quenched disorder, the microemulsions are equilibrium systems where interactions influence the formation of a percolation cluster [3]. In addition, the percolation clusters consisting of water droplets are not static. They are capable of rearranging through diffusion in the surrounding medium.
These differences led to the conjecture [4] that these systems display a different kind of percolation, called the "stirred" percolation. In this paper we describe experimental results that demonstrate the differences between percolation transition in microemulsions and that in the standard systems. The system investigated [5] is a decane/water/aerosol OT (sodium di-2-ethyl hexyl sulfosuccinate) microemulsion with a water and surfactant volume fraction of 8%. The molar ratio of water to surfactant is 40.8. This system has been studied extensively in terms of its structure [6] and the phase separation transition. Neutron scattering data show that the system consists of surfactant-covered water droplets of 50 Å-radius.

Since interactions play an important role in the percolation process, the percolation transition can be accessed not only by varying the concentration ϕ of the water-surfactant, but also by varying temperature. This is an unusual feature of the microemulsion system. Fig. 1 shows the temperature dependence of the conductivity σ and the dielectric constant ε at two frequencies. The phase separation transition is marked by T_c. Both ε and σ increase as T is increased. ε reaches a peak while σ continues to rise. Strong frequency dependence is also observed near the peak in ε, but decreases on either side. The system shows all the features of a poor conductor-good conductor percolation transition.

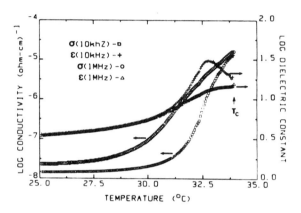

Fig. 1. Temperature dependence of the conductivity and the dielectric constant at two frequency. T_c marks the phase-separation transition.

Recent theoretical studies [7] have revealed the importance of interactions among the droplets in forming the percolation cluster. If ϕ_p, the percolation threshold, is analytic in T, then one expects to find the same critical exponent by varying T as by varying ϕ which is the common procedure.

For a poor conductor-to-good conductor percolation transition the metal-superconductor (MS) analogy for σ is relevant below threshold and an insulator-metal (IM) analogy is relevant above threshold. Thus, one obtains

$$\sigma \sim (T_p - T)^{-s'} \quad \text{for } T < T_p \tag{1}$$

$$\sim (T - T_p)^{\mu'} \quad \text{for } T > T_p$$

with a cross-over regime near T_p whose width depends on the good conductor-to-poor conductor conductivity ratio. We obtained the value of T_p by locating the peak in $\frac{1}{\sigma} d\sigma/dt$ which coincides with the peak in ε at low frequencies.

Fig. 2(a) shows the fit of σ to the forms given in Eqn. (1) above and below T_p. The scaling regime yields exponents $\mu' = 1.68 \pm 0.05$ and $s' = 1.17 \pm 0.05$. Significantly, these exponents are the same as those obtained in the same system [8] by varying ϕ at a fixed T. Moreover, they are in excellent agreement with results obtained in other microemulsions systems [4]. This suggests that the scaling behavior is generic in microemulsions.

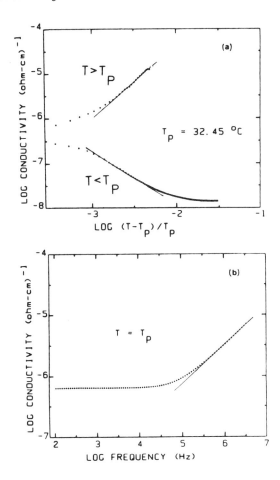

Fig. 2. (a) Scaling behavior of the low frequency conductivity below and above the percolation threshold.
(b) Frequency scaling of the conductivity near T_p. The crossover from frequency dependent to frequency independent regimes marks the effect of rearrangement.

Comparison of the experimental results to theoretical expectations leads to an interesting situation. The measured value of μ' is in excellent agreement with the IM analogy where $\mu \sim 1.7-2.0$ for $d = 3$. But s' is in serious disagreement with the MS analogy where $s \sim 0.6-0.7$ for $d = 3$. This implies that the assumption of a static percolation inherent in evaluating s is not applicable here.

Recently a theoretical model [7] has been developed in order to understand the role of the mobility of the water droplets. In this model, the charge transport proceeds via the hopping of surfactants on the droplets which diffuse resulting in a rearrangement of clusters. This results in an increase in the conductivity below the percolation threshold. At low frequencies, the theory predicts the following form of σ

$$\sigma \sim (T_p - T)^{-\hat{s}} \; ; \; \hat{s} = 2\nu - \beta \tag{2}$$

where ν is the exponent of the correlation length and β describes the probability of being on the infinite cluster. For $d = 3$, $\hat{s} \sim 1.3$ which is in agreement with the measured value. In this model, the exponent μ above threshold remains unaltered.

The dynamic nature of percolation leads to interesting frequency dependence of σ, also in contrast with static percolation. The overall scaling form is given by

$$\sigma(\omega, T_p - T) \sim (T_p - T)^{-(2\nu - \beta)} f(\omega \tau_R, \omega \tau) \tag{3}$$

where τ_R is the rearrangement time and $\tau \sim \xi^{2+\theta}$ represents the anomalous diffusion time on the cluster[9], where $2 + \theta = (\hat{s} + \mu)/\nu$. At the percolation threshold two different frequency regimes are expected: (1) $\omega < 1/\tau_R$ where σ is independent of frequency and (2) $\omega > 1/\tau_R$ where σ is of the form

$$\sigma \sim \omega^x \; ; \; x = \frac{\mu}{\hat{s} + \mu} \tag{4}$$

Experimental results on the frequency dependence of σ at $T \approx T_p$ is shown in Fig. 2(b). Indeed, there is a frequency independent regime below $f = 30$ KHz. Above this frequency there is a power-law regime of mean slope 0.63 ± 0.04. Within the accuracy of the measurement, this is in excellent agreement with equation (4) with the measured values of $\hat{s}$ and μ.

Furthermore, from the onset of a power-law regime we obtain an estimate of the rearrangement time $\tau_R \sim 5$ μsec. This number is in excellent agreement with the self-diffusion time for a droplet of radius 50 Å given by the Stokes-Einstein formula

$$\tau = \frac{6\pi R^3 \eta}{k_B T} \tag{5}$$

where R is the radius of the droplet and η is the viscosity of the system. Thus the experimental results on the conductivity are in excellent agreement with the dynamic percolation model.

The results on the dielectric constant are less clear. The primary problem is the small range of variation of ε. Whether the asymptotic scaling regime is ever reached in this experiment is not clear. The backbground subtraction to eliminate the effect of oil can result in a significant alteration of the effective critical exponent. However, recent high frequency measurements [10] demonstrate that at high frequencies ε obeys a scaling relation at T_p

$$\varepsilon \sim \omega^{-y} \tag{6}$$

where y = 1-x consistent with the values obtained here. The difference between the measured value and the value expected for static percolation (s replacing $\hat{s}$ in Eqn. (4)) again confirms the dynamic percolation picture.

To conclude, percolation in microemulsions where the charge carriers hop on a dynamically rearranging cluster provides a rich and interesting set of phenomena not commonly observed.

We acknowledge helpful discussions with S. Alexander, A. Bug, G. Grest, S. Safran and I. Webman.

REFERENCES

1. For a recent review of percolation, see <u>Percolation Structures and Processes</u>, Edited by G. Deutscher, R. Zallen and J. Adler, Annals of the Israel Physical Society, Vol. 5 (Israel Physical Society, Jerusalem, Israel, 1983).
2. A. M. Cazabat, D. Chatenay, F. Guering, D. Langevin, J. Mennier, O. Sorba, J. Lang, R. Zana and M. Pailette in <u>Surfactants in Solution</u>, Edited by K. L. Mittal and B. Lindman (Plenum, New York, 1984), p. 1137 and references therein.
3. S. A. Safran, I. Webman and G. S. Grest, Phys. Rev. A <u>32</u>, 506 (1985).
4. M. Lagües, J. Paris (Paris), Lett. <u>40</u>, L-331 (1979).
5. S. Bhattacharya, J. P. Stokes, M. W. Kim and J. S. Huang, Phys. Rev. Lett. <u>55</u>, 1884 (1985).
6. J. S. Huang, S. A. Safran, M. W. Kim, G. S. Grest, M. Kotlarchyk and N. Quirke, P hys. Rev. Lett. <u>53</u>, 592 (1984) and references therein.
7. G. S. Grest, I. Webman, S. A. Safran and A.L.R. Bug, Phys. Rev. A<u>32</u>, 2842 (1986).
8. M. W. Kim and J. S. Huang, Phys. Rev. A<u>34</u>, 719 (1986).
9. Y. Gefen, A. Aharony and S. Alexander, Phys. Rev. Lett. <u>50</u>, 77 (1983).
10. M. Van Dijk, Phys. Rev. Lett., <u>55</u>, 1003 (1985).

FRACTAL SURFACES IN POROUS MEDIA

Po-zen Wong
Schlumberger-Doll Research, Old Quarry Road, Ridgefield, CT 06877-4108

ABSTRACT

Recent studies have shown that porous materials often have fractal internal surfaces. In the article, we give a brief review of what fractal surfaces are, how they are observed experimentally, why they form, and how they affect the a.c. electrical transport properties through the media. Theoretical results are compared to real systems such as sedimentary rocks and rough/porous electrodes.

I. INTRODUCTION

Porous media can generally be considered as two-phase systems consisting of solids and pores separated by random surfaces. Historically, in studying the physical properties of porous materials, there is a tendency to focus on either the solid or the pores. For examples, elastic properties are mostly influenced by the degree of continuity of the solid,[1] and transport properties are usually thought to be controlled by the connectivity of the pores.[2] Relatively little attention has been paid to the interface that divides the solid and the pores. In fact, looking back at the Proceedings of the Symposium held here in 1983, there was not a single paper related to the pore-solid interface. During the last several years, however, we have seen increasing interests and efforts to charaterize these surfaces and to understand their effects on the bulk properties of porous materials. The notion that surfaces can affect bulk properties may seem unusual to some of us who are accustomed to thinking about an infinite solid when we studied solid-state physics in school, but the reason here is quite simple. If we think of porous materials as composites of small particles, such as sandstones are made up of sand grains, it is clear that a large fraction of the atoms in the system can be on the surface. This fraction is roughly given by $f \sim aS/V \sim a/R$, where S/V is the surface to volume ratio, a is the lattice constant, and R is the particle size. Hence, if one studies these systems by statistical mechanics, it is neccessary to keep in mind the surface atoms. Conventional results of condensed matter physics can break down in dealing with porous systems with micro-particles and micro-pores.

To give a simple example, we recall the Einstein's relation

$$\sigma_w \propto nD_c, \qquad (1.1)$$

which says that the conductivity σ_w of an aqueous solution is proportional to the ionic concentration n and the ions' diffusion constant D_c. If one saturates the pores of a rock with this solution, one would expect the rock conductivity σ_r to be proportional to a reduced concentration $n\phi$, where ϕ is the porosity, and a reduced effective diffuion constant $D_c{}'$, i.e.,

$$\sigma_r = n\phi D_c' = \sigma_w \phi D_c'/D_c \equiv \sigma_w \phi/\alpha_T \ . \tag{1.2}$$

The ratio $\alpha_T \equiv D_c/D_c'$ reflects how diffusion is slowed down by the tortuous paths through the pore and it is called the *tortuosity*.[3] (In the well known empirical relationship due to Archie,[4,2] α_T is written as ϕ^{1-m}. To my knowledge, this has not been proven from a diffusion point of view). From Eq. (1.2), one expects σ_r to be directly proportional to σ_w. However, Waxman and Smit[5] have found in 1968 that this is not true in sandstones. Fig. (1) shows recent data obtained in our laboratory. It shows a nonlinear dependence of σ_r on σ_w for NaCl solution, but a linear dependence for $(C_2H_5)_4NCl$ solution. This result shows that the bulk conductivity is modified by interactions between the ions in the pore fluid and the solid surfaces, but the details are not understood at present. We suspect that similar phenomena can occur to the viscosity of fluids. Putting the fluid inside a porous material might lead to a different effective viscosity because of the abundance of interactions between the fluid molecules and the solid surfaces.

Any study of surface effects cannot be complete without knowing something about the geometry of the surface, e.g., whether it is smooth or rough, how the roughness is characterized, etc.. During the last three years, there has been increasing experimental evidence that many porous materials have *fractal* internal surfaces. Some examples are: Avnir et al's work using absorption isotherms,[6] Katz and Thompson's work using microscopy,[7] Bale and Schmidt's work using X-ray scattering[8] and my own work with Howard and Lin using neutron scattering.[9] In the following, I give a brief description of what *fractals* surfaces are, how they are observed by experimentally, why they exist, and how they affect the ionic transport through the porous media. Experimental examples includes sedimentary rocks and rough/porous electrodes.

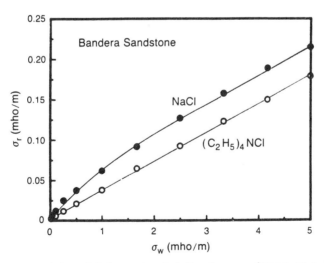

Fig. 1: The conductivity of brine saturated Bandera sandstone as a function of brine conductivity. The solid circles are for NaCl solution and the open circles for $(C_2H_5)_4$ solution.

II. FRACTAL SURFACES: SELF-SIMILAR AND SELF-AFFINE

Fractal is a term used to describe geometrical objects which are invariant upon the change of length scales,[10] i.e., these objects have geometrical features over a wide range of sizes and they look similar on all length scales. The concept of scale invariance, or *scaling*, has had profound impact in modern condensed matter physics, especially in the context of second-order phase transition.[11] In applying this mathematical concept to real systems, it is important to remember that there usually exist upper and lower limits for the size scale over which the system is self-similar. For second-order phase transitions, the lower limit is the lattice constant and the upper limit is the correlation length ξ.

A simple example of a fractal is the so-called Sierpinski gasket, illustrated in Fig. 2a. This object is produced by starting from a black triangle of size L and cutting out successively smaller triangular holes (in white) of size $L/2^n$, where $n=1,2,3,...$. When observed with a length resolution r, one does not see the holes that are smaller than r and the black triangles of that size appear solid. The number of such triangles increases as a power of the ratio L/r:

$$N_r \sim \left(\frac{L}{r}\right)^D , \qquad (2.1)$$

where D is called the Hausdorff dimension, or the fractal dimension. D can have any value between 0 and the dimension d of the Euclidean space in which the object is embedded in. For the Sierpinski gasket, $D=\log3/\log2 \approx 1.584962...$.

If one measures the area of the black triangles in Fig. 2a with a resolution r, one finds

$$A_r = \frac{1}{2}r^2 N_r \sim r^{2-D} L^D \qquad (2.2)$$

The average areal density is $A_r/L^2 \sim (r/L)^{2-D}$, which goes to zero as $r/L \to 0$. Fractal objects with this type of behavior can be called called *mass fractals*. In contrary, the collection of white triangles in Fig. 2a are not mass fractals because they have finite area and areal density as $r/L \to 0$. However, if we measure the length l_r of the perimeter of the white triangles with a resolution r, we find

$$l_r = 3r \cdot \frac{1}{3} N_r \sim r \left(\frac{L}{r}\right)^D \to \infty \quad \text{as } r/L \to 0 . \qquad (2.3)$$

Fractals with this behavior can be called *boundary fractals*. For spatial dimension $d=3$, a mass fractal can be called a *volume fractal*, and a boundary fractal can be called a *surface fractal*.

Fractal objects we encounter in nature are generally *random*, not created by *deterministic* rules like the Sierpinski gasket. Fig. 2b shows an example of random fractals. It is a plot of the displacement (Y) of a one-dimensional random walker versus time (X). Since the vertical and horizontal directions are inequivalent, this line is *anisotropic*, showing only vertical fluctuations. This is an example of *self-affine* fractals,[12] as opposed to the Sierpinski gasket which is *self-similar*. Self-

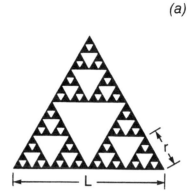

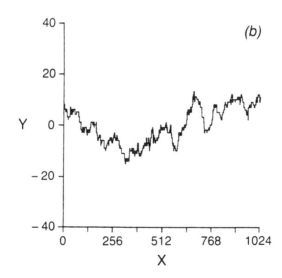

Fig. 2: (a) The Sierpinki gasket is an example of a deterministic self-similar fractal. The black triangles form a mass fractal and the white ones has only fractal boundary. (b) The displacement Y of a one-dimensional random walker is plotted against time X. The line is an example of a random self-affine fractal with $\alpha = \frac{1}{2}$.

similar fractals rescale the same way along any spatial direction, but self-affine fractals require different rescaling factors for the horizontal and vertical directions. For the line in Fig. 2b, we know that $<Y^2> \propto X$ for random walks, so if $X \to pX$ and $Y \to qY$, then $q = p^{1/2}$. In general, self-affine fractals have $q = p^\alpha$, with α between 0 and 1. This means that the vertical root-mean-square fluctuation is given by

$$w \equiv <Y^2>^{1/2} \propto X^\alpha \qquad (2.4)$$

The fractal dimension of the line can be obtained by the following consideration. First, we note that in plotting the Y versus X on a graph paper requires choosing phsycial units of length and time for the two axes. Once the choice is made, the position of every point on the line is measured by the number of horizontal and vertical grids from the origin. We treat the grid size as the length unit. We then introduce another characteristic length b (which is also measured in number of grids) and rewrite Eq. (2.4) as

$$w = b(X/b)^\alpha \qquad (2.5)$$

b is a special length because if $X = b$ then $w = X$. We call b the crossover length.[9] The reason is that if we measure the length of the line l_r with a resolution r over a horizontal distance L, then there are L/r horizontal segments and each has a vertical fluctuation of order w, so

$$l_r = \frac{L}{r}(r+w) = L[1+(b/r)^{1-\alpha}] \qquad (2.6)$$

Since $1-\alpha > 0$, we have $l_r \approx L$ for $r \gg b$ and the line is effectively smooth for large r. However, for $r \ll b$, we have $l_r \approx L(b/r)^{1-\alpha} \to \infty$, which indicates fractal behavior. Comparing the exponent of r to that in Eq. (2.3), we identify $D = 2-\alpha$. For a general spatial dimension d, we have

$$D = d - \alpha \qquad (2.7)$$

for self-affine boundaries. Self-affine surfaces in three dimensions has $D = 3-\alpha$.

III. MEASURING FRACTAL DIMENSIONS

Since fractal structures in real systems are generally random, their dimensions cannot be calculated like the Sierpinski gasket and must be measured experimentally. This can be done either in *real* space by some size measurement techniques, or in *reciprocal* space by scattering techniques.

Avnir et al studied fractal surfaces of porous materials by molecular absorption technique using molecules of different sizes.[6] This corresponds to varying r in Eq. (2.2) and it probes a length scale between a few angstroms and several tens of angstroms. Brown and Scholz studied self-affine surfaces of rocks between 10 μm and 1 m by measuring the vertical height profiles,[13] like Y versus X in Fig. 2b. They calculated the height-height correlation function $G(r) \equiv <(Y_o - \bar{Y})(Y_r - \bar{Y})>_o \equiv w^2$ along a horizontal straight line and its Fourier transform $\tilde{G}(q)$. The exponent α was obtained by using either

$$G(r) \sim w^2 \sim r^{2\alpha} \tag{3.1}$$

or

$$\tilde{G}(q) \sim q^{-(1+2\alpha)} \tag{3.2}$$

The fractal dimension is obtained from Eq. (2.7) with $d=2$. Katz and Thompson studied fractal surfaces in sandstones using microscopy techniques which covers a length scale of 0.1 to 10 μm.[7] They obtained D by determining the size distribution of "features" along straight lines across the pictures. This is possible because a fractal surface can be considered to have a power-law distribution of feature sizes, i.e.,[14]

$$\frac{dN}{dr} \sim r^{-\beta}, \quad \text{where } \beta = D+1 \tag{3.3}$$

If one draws a straight line to cut across this surface, the probability of cutting through a feature of size r is proportional to r^3, so the features size distribution along the line is

$$\frac{dN_1}{dr} \sim r^{3-\beta} \sim r^{2-D} \tag{3.4}$$

In each of the above studies, there is ample evidence for fractal behavior but sometimes it is difficult to distinguish whether it is a volume fractal or a surface fractal, whether it is self-affine or self-similar etc.. To some extent, these ambiguities can be alleviated by scattering experiments.

If we define the local density $\rho(x)$ in a porous material to be zero in the pore and one in the solid, then according to the Born approximation, the scattering cross-section is simply proportional to the Fourier transform of the density-density correlation function, i.e., the intensity I as a function of wavevector q is given by

$$I(\mathbf{q}) \propto \int g(r) e^{i\mathbf{q} \cdot \mathbf{r}} d^d\mathbf{r} \tag{3.5}$$

where $g(r) \equiv \langle \rho_0 \rho_r \rangle$. Debye et al[15] and Porod[16] have shown that for small r,

$$g(r) \approx 1 - \frac{rS}{4\phi(1-\phi)V} \tag{3.6}$$

where S/V is the surface to volume ratio. The leading term of unity gives a $\delta(\mathbf{q})$ term in $I(\mathbf{q})$. This is not measured experimentally because it coincides with the incident beam, and it shall be neglected in the following. The term linear in r gives a q^{-4} behavior in the small-angle scattering which is known as the Debye-Porod law. It applies to systems with random surfaces which are smooth on the length scales of q^{-1}, i.e., the area S in Eq. (3.6) is independent of r ($r \equiv q^{-1}$).[14] For a fractal surface,

$$S \sim r^{2-D} \tag{3.7}$$

Putting this into Eqs. (3.6) and (3.7) gives

$$I(q) \sim \frac{1}{q^{6-D}} \tag{3.8}$$

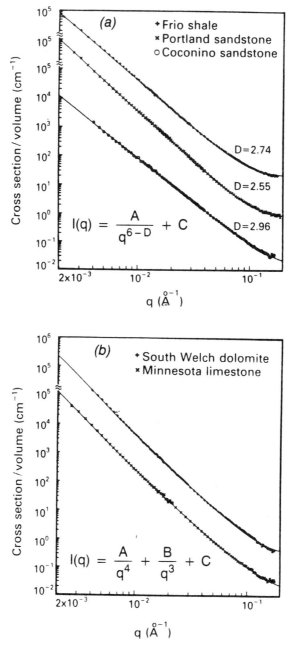

Fig. 3: (a) Small-angle neutron scattering data on sandstones and shales are described by a $1/q^{6-D}$ behavior plus an incoherent scattering background, consistent with having fractal surface. (b) The data on carbonate rocks are well fitted by Eq. (3.9) for self-affine surfaces.

This behavior has been observed by Bale and Schmidt in lignite coals using small-angle X-ray scattering.[8] Fig. 3a shows similar results in sandstones and shales, obtained by myself with Howard and Lin using small-angle neutron scattering.[9] For self-affine surfaces, because

$$S \sim S_o(1+\frac{w}{r}) \sim S_o[1+(b/r)^{1-\alpha}]$$

substituting into Eqs. (3.6) and (3.5) gives[14]

$$I(q) \sim \frac{A}{q^4} + \frac{B}{q^{3+\alpha}} \tag{3.9}$$

where $B/A = b^{1-\alpha}$ determines the crossover length b. Since $\alpha = 3 - D$, the $q^{-(3+\alpha)}$ term in Eq. (3.9) is the same as the $q^{-(6-D)}$ term in Eq. (3.8). The additional q^{-4} term in Eq. (3.9) reflects the fact that self-affine surfaces becomes effectively smooth at large length scales and the behavior must approach the Debye-Porod law in that limit. Fig. 3b shows small-angle neutron scattering data on limestones,[14] which can be fitted to Eq. (3.9) with $\alpha = 0$, consistent with self-affine behavior. This is a plausible interpretation at present and it requires confirmation by further work.

One advantage of scattering experiments is that they can readily distinguish volume fractals from surface fractals. Volume (or mass) fractals have $g(r) \approx 1 - r^{D-3}$. From Eq. (3.5), this gives a small-angle scattering intensity

$$I(q) \sim \frac{1}{q^D} \tag{3.10}$$

where the exponent D is less than 3. This contrast with Eq. (3.8) for surface fractals, where the exponent $6-D$ is greater than 3.[14] Another advantage of scattering is the efficiency in data collection as compared to the real space measurements. However, we should note that scattering experiments are only suitable for probing sub-micron structures. State-of-the-art high resolution synchrotron X-ray scattering[17] can only measure features below about 5000Å. Thus large scale structures must rely on real space measurements.

IV. SURFACE TENSION OF FRACTAL SURFACES

Why do surfaces become fractals? On a qualitative level, one suspects that there must be some physical mechanisms that favors the surface to fluctuate rather than to stay flat. These fluctuations increases the surface area relative to a smooth surface. In thermodynamics language, we can say that the surface tension is negative. This idea has recently been discussed by Cohen and Anderson.[18] Here, I give a specific example which is drawn from the solid-on-solid (self-affine) interface model developed by Grinstein and Ma for the random-field Ising model (RFIM).[19]

The RFIM has a set of Ising spins placed on a d-dimensional lattice. The Hamiltonian of the model is

$$\mathbf{H} = -\sum_{<i,j>} JS_iS_j - \sum_i H_iS_i \qquad (4.1)$$

where S_i is the spin variable at the ith site and it can have values ± 1, which represents the spin pointing up or down. The first term in Eq. (4.1) shows that the spins on neighboring sites i and j interact with an exchange energy J which favors the spins to be parallel. The second term in adds a magnetic field H_i which points up and down randomly at every site with $<H_i>=0$ and $<H_i^2>=\Delta$. The two interactions compete with each other. The exchange term favors the spins to all point in the same direction, and the random-field term favors the spins to point up and down randomly. In order to determine whether the system would have long-range order, Grinstein and Ma assumed that an interface exists between a spin-up domain and a spin-down domain and considered its equilibrium shape using the self-affine geometry in Fig. 2b. If w is the vertical fluctuation over a horizontal length r, the number of sites involved is $r^{d-1}w$. Statistically, the random field energy summed over these sites is

$$E_H = (r^{d-1}w\Delta)^{1/2} \qquad (4.2)$$

The surface fluctuation lowers the system's energy by this amount. On the other hand, the surface breaks the exchange bonds and costs an energy

$$E_J = Jr^{d-1}(1+\frac{w}{r}) \qquad (4.3)$$

The net surface energy is $E_S = E_J - E_H$. The equilibrium value for w is determined by $\partial E_S/\partial w = 0$, which gives

$$w = \frac{\Delta}{4J^2}r^{3-d} \qquad (4.4)$$

This implies $\alpha = 3-d$ in Eq. (2.5), and $D = 2d-3$ by Eq. (2.7). Substituting w back into Eqs. (4.2) and (4.3), we find

$$E_S = Jr^{d-1}(1 - \frac{\Delta}{4J^2}r^{2-d}) \qquad (4.5)$$

The surface tension is simply the surface energy per unit area

$$T = E_S/r^{d-1} = J(1 - w/r) \qquad (4.6)$$

Hence, the surface tension is negative for $w > r$, and positive for $w < r$. These conditions are exactly the same as those we used for deriving the fractal dimension from Eq. (2.6). This example shows explicitly the correspondence between negative surface tension and the fractal character of a surface.

The RFIM is in many ways similar to porous media. The exchange energy is like the chemical bonds between atoms in the solid. The random field are similar to impurity atoms that cost extra energy to incorporate into the solid. The solid surface like to bend around to exclude as many impurities as possible, similar to the domain walls that like to exclude sites with random fields pointing in the wrong direction. Of course, these analogies are not exact, but the idea that competing interactions lead to negative surface tension is of fundamental importance in understanding the origin of fractal surfaces in porous media. In our neutron scattering studies of sandstones, we found that the fractal surfaces are due to the presence of clays.[9] These compounds typically contain a large amount of impurities and there are local charged sites which have to be kept neutral by chemi-absorption of cations on the surface. These interactions favor a high surface area. On the other hand, there are neutral sites in the clay which prefer not to have surfaces because it cost energy to break bonds at the surface. This is another example of competing interactions. The next step of understanding will be to see how crystal growth mechanisms or details of the interactions might lead to different fractal dimensions. The RFIM gives a unique dimension $D=2d-3$. It does not explain why a wide range of D's is observed in different materials.

V. A.C. RESPONSE OF FRACTAL SURFACES

Knowing that some material has a fractal structure is not of much value unless we also understand how it affects the physical properties of that material. In this section, we consider the frequency dependence in the conductivity and dielectric dispersions, $\sigma(\omega)$ and $\epsilon(\omega)$, of brine saturated rocks. The rocks themselves are insulating, but the ions in the pore water conduct. When a electric field is applied, the ions move along the field and get blocked by the solid surfaces. This is like charging up a capacitor. A rock with fractal surfaces is like a network of resistances and capacitors, but the capacitor plates have rough surfaces.

Let us first consider the a.c. impedance of a special parallel-plate capacitor system, namely, an electrochemical cell consisting of an aqeous electrolyte and two *blocking* electrodes. By *blocking*, we mean the electrodes are inert, they do not react with the electrolyt chemically and no charge can be transported across the interface. The impedance at low frequency is, of course,

$$Z^*(\omega) = \frac{1}{i\omega C} = \frac{R}{i\omega \tau_w} \tag{5.1}$$

where R and C are the resistance and capacitance of the electrolyte between the electrodes, $\tau_w = RC = \rho_w \epsilon_w$ is the the relaxation time of the electrolyte. For a NaCl solution with resistivity $\rho_w = 1\,\Omega$-m and $\epsilon_w \approx 80\epsilon_o$, $\tau_w \approx 7\times 10^{-10}$ sec. For $\omega\tau_w \ll 1$, Eq. (5.1) is valid, because the space charge layer at the interface in fully developed for time $t \gg \tau_w$. For short times ($\omega\tau > 1$), however, we have to consider how the charge layer is built up. Assuming that the electric field is switched on like a step function at $t=0$, then for time $t \ll \tau_w$, the thickness of the charge layer λ must increase according to the diffusion relation $\lambda^2 = D_c t$, where D_c is the diffusion constant. The reason is that if there were no diffusion, the charge at the interface would be infintesimally thin. Diffusion spreads out this concentration gradient. Over a time t, the charge layer must spread out to an thickness given by the diffusion length λ. In fact, the equilirium thickness λ_D of the layer, i.e., the Debye-Hückle screening length, is just the diffusion length for time τ_w. With this understanding, we can see that the polarization developed in response to the electric field is

$$P = n\lambda \propto t^{1/2} \propto \omega^{-1/2} \qquad (5.2)$$

where n is the charge density. Similarly, the average velocity v for time t is λ/t and the conductivity is

$$\sigma = nv \propto t^{-1/2} \propto \omega^{1/2} \qquad (5.3)$$

These give a capacitance $C \sim \omega^{-1/2}$ and a resistance $R \sim \omega^{-1/2}$, so the impedance for $\omega \tau_w \gg 1$ is

$$Z^*(\omega) \propto R(i\omega \tau_w)^{-1/2} . \qquad (5.4)$$

This is known as the Warburg impedance in electrochemistry.[20]

In general, it is not neccessary to have perfectly flat electrodes to observe the Warburg impedance. Chew and Sen[21] have studied the dielectric function of a charged sphere suspended in an electrolyte. They solved the problem analytically and the solution contains a term of the form $(i\omega\tau)^{1/2}$, but τ here is the diffusion time associated with the radius a of the sphere, i.e., $a^2 = D_c \tau$. The physical reason is that the charges moving along the field are blocked by the sphere. For them to move passed the sphere, they must first diffuse perpendicular to the field direction. Since diffusion is much slower than the drift along the field, the controlling time for the system to equilibrate is the diffusion time τ. Along the same line or arguments, if the electrodes in the previous example were not perfectly flat, but have some micro-features of size a, the frequency range for observing the Warburg impedance would become $\omega\tau \gg 1$, not $\omega\tau_w \gg 1$. For $a = 10\mu$m, one only needs $\omega \gtrsim 15$ Hz. This is probably why the phenomenon is observable in electrochemical impedance measurements.

For fractal electrode surfaces, the above arguments can be extended quite easily. Consider the self-affine geometry in Fig. 2b with an electric field applied in the vertical direction. As the ions reach the surface, they are blocked. The only way to move forward further is to diffuse in the horizontal direction and then drift along the field vertically. For a horizontal diffusion distance r, the average vertical displacement is $w \sim r^\alpha$. The polarization developed in time t is

$$P \sim nw \sim r^\alpha \sim r^{d-D} \sim t^{(d-D)/2} \sim \omega^{(D-d)/2} \qquad (5.5)$$

Hence, the capapcitance C should also have the same frequency dependence. The exponent $\dfrac{D-d}{2}$ is between zero and $-\tfrac{1}{2}$. According to Scheider,[22] fractional power-law behavior with this range of exponent is commonly observed in electrochemical cells with liquid electrolytes. Recently, Bates et al[23] have reported the same for solid electrolytes. Similar to Eq. (5.3), the conductivity is proportional to the average velocity w/t, so

$$\sigma \sim nw/t \sim \omega^{1 + \frac{D-d}{2}} \qquad (5.6)$$

Thus, we can write down the frequency dependence of the impedance as

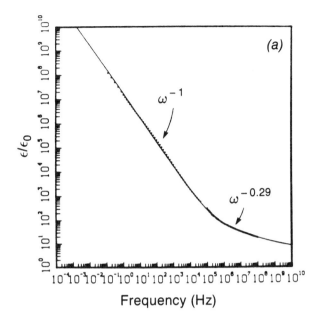

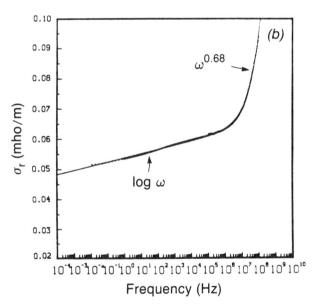

Fig. 4: (a) Dielectric dispersion of a Portland sandstone saturated with 1 Ω-m NaCl solution. It shows a $1/\omega$ behavior at low frequency and an $1/\omega^{0.29}$ behavior at high frequency. (b) The conductivity dispersion shows a $\log\omega$ behavior at low frequency and an $\omega^{0.68}$ behavior at high frequency. The high frequency results are consistent with Eqs. (5.7) and (5.8).

$$Z^*(\omega) \sim (i\omega)^{-z} \tag{5.7}$$

where $z = 1 + \frac{D-d}{2}$ is between ½ and 1. For smooth surfaces with $D = d - 1$, these results recover Eqs. (5.2)-(5.4).

For porous materials in three dimensions, we should have

$$\epsilon \sim \omega^{-\frac{3-D}{2}} \tag{5.7}$$

and

$$\sigma \sim \omega^{\frac{D-1}{2}} \tag{5.8}$$

These results can be obtained without the assumption of a self-affine surface. As stated in Eq. (3.3), a fractal surface has a power-law distribution in feature size. The polarization due to a feature of size r (a sphere for example) is proportional to the surface charge Q_s, which is proportional to r^2, times the charge separation r, with some relaxation function $f(\omega\tau)$. The dielectric function for the surface is just the sum of the response due to all the features. Using Eq. (3.3), we have

$$\epsilon \sim \int r^3 f(\omega\tau) r^{-(D+1)} dr \tag{5.10}$$

Since $\tau \sim r^2$, counting the powers of r in the integral immediately yields Eq. (5.8). Combining Eqs. (5.8) and (5.9) gives a complex dielectric function

$$\epsilon^* \sim (i\omega)^{-\frac{3-D}{2}} \tag{5.11}$$

Experimental tests of the above predictions have been made on a number of sandstones in our laboratory. Fig. 4a shows the dielectric dispersion of a Portland sandstone saturated with 1 Ω-m NaCl solution. There is a $1/\omega$ behavior in the low frequency region which is common to all shaly sandstones and it is due to the cation exchange chemistry between the clay and the solution. The high frequency tail in Fig. 4a indicates an ω^{-x} behavior, with $x = 0.29$. The fractal dimension D for this rock was determined by neutron scattering and we found $D = 2.55$ (see Fig. 3a). The predicted exponent for the dielectric dispersion is $x = (3-D)/2 = 0.23$. Similarly, we show in Fig. 4b the conductivity dispersion for the same rock. There is an ω^y behavior at high frequency, with $y = 0.68$. The theoretical prediction is $y = (D-1)/2 = 0.77$. In general, we find that the electrical measurements and the neutron measurements for x and y are consistent to about 0.1. The theoretical bounds, $0 \leq x \leq ½$ and $½ \leq y \leq 1$, are always obeyed. In addition, we note that Eq. (5.11) requires $x + y = 1$ and this is always true to within experimental error. Finally, the frequency range for observing the fractional power-laws are related to a size range through the diffusion relation $r^2 = D_c \tau$. Using $D_c = 1.5 \times 10^{-5}$ cm^2/sec and $\tau = 1/2\pi f$, we find that $f = 10^9$ Hz corresponds to $r = 4$Å and $f = 10^5$ Hz corresponds to 400 Å. These coincide with the range of length scales probed by neutron scattering. A similar comparison in electrode systems are more difficult, because the impedance measurements are generally made at lower frequencies which often corresponds to structures in the 0.1–10 μm range.

VI. SUMMARY

Surface effects are important in any porous material which has a high surface to volume ratio. This is a theme shared by several speakers at this conference. In this paper, I have focused on the fractal aspects of the pore-solid interface, discussed the difference between self-affine and self-similar fractals and how they are measured experimentally. The random-field Ising model is used to demonstrate how competing interactions can lead to negative surface tensions on a microscopic scale and produce the fractal character. Of course, for large scale fractals such as the surface of the earth or the fractured surface of rocks, purely random Brownian processes (like the random walk) would give $D = d - \frac{1}{2}$.[12,13] I have used the a.c. response of fractal surfaces in the context of electrochemical cells and brine saturated rocks as examples to show how these surfaces affect the bulk properties of porous materials. My results are by no means definitive. There have been numerous theoretical attempts to treat this problem[22,24-28] but the results obtained by different author groups are all different. Perhaps unsurprising, they also differ from the expressions in Eqs. (5.7) and (5.8) which were derived by heuristic arguments. To bias the readers, I should point out that the two exponents in those equations were also obtained by de Gennes[29] in a somewhat different context. The experimental data we have on rocks at present are in good agreement with these theoretical predictions! I hope more precise results will be obtained in the near future.

ACKNOWLEDGEMENTS

The neutron scattering data are obtained in collaboration with J.-s. Lin of Oak Ridge National Laboratory, and the electrical data with J. P. Tomanic. I owe much to both of them. I am grateful to my colleagues J. Howard and N. Wada for helpful discussions on the properties of clay, and to L. Schwartz for providing Fig. 2b. I wish to also thank all my colleagues at Schlumberger who have stimulated my interest in porous media and fractals.

After the completion of this manuscript, I learned that some of my discussion on self-affine fractals had been given independently by Dr. S. Alexander.[30] I thank him for sending me a copy of his preprint.

REFERENCES

1. See, e.g., L. Schwartz, in *AIP Conference Proceedings No. 107, Physics and Chemistry of Porous Media*, edited by D. L. Johnson and P. N. Sen, (American Institute of Physics, 1984), p.105.

2. P.-z. Wong, J. Koplik, and J. P. Tomanic, Phys. Rev. B 30, 6066 (1984).

3. D. L. Johnson, T. J. Plona, C. Scala, F. Pasierb, and H. Kojima, Phys. Rev. Lett. 49, 1840 (1982).

4. G. E. Archie, AIME Trans. 146, 54 (1942).

5. M. H. Waxman and L. J. M. Smit, AIME Trans. 243, 107 (1968).

6. D. Avnir, D. Farin and P. Pfeifer, Nature 308, 261 (1984).

7. A. J. Katz and A. H. Thompson, Phys. Rev. Lett. 54, 1325 (1985).

8. H. D. Bale and P.W. Schmidt, Phys. Rev. Lett. 53, 596 (1984).

9. P.-z. Wong, J. Howard, and J.-s. Lin, Phys. Rev. Lett. 57, 637 (1986).

10. See, e.g., B. B. Mendelbrot, *The Fractal Geometry of Nature* (Freeman, San Francisco, 1982).

11. See, e.g., L. P. Kadonoff et al, Rev. Mod. Phys. 39, 395 (1967).

12. See, e.g., R. F. Voss in *Scaling Phenonmena in Disordered Systems*, edited by R. Pynn and A. Skjeltorp (Plenum, New York, 1985).

13. S. R. Brown and C. H. Scholz, J. Geophys. Res. 90, 12575 (1985).

14. P.-z. Wong, Phys. Rev. B 32, 7417 (1985).

15. P. Debye, H. R. Anderson and H. Brumberger, J. Appl. Phys. 28, 679 (1957).

16. G. Porod, in *Small Angle X-Ray Scattering*, edited by O. Glatter and O. Kratky (academic, New York, 1982).

17. See, e.g., P. Dimon et al, Phys. Rev. Lett. 57, 595 (1986).

18. M. H. Cohen and M. P. Anderson, in *The Chemistry and Physics of Composite Media*, edited by M. Tomkiewicz and P. N. Sen (The Electrochemical Society, Pennington, New Jersey, 1985).

19. G. Grinstein and S.-k. Ma, Phys. Rev. B 28, 2588 (1983).

20. E. Warburg, Ann. Phys. Chem. 67, 493 (1899).

21. W. C. Chew and P. N. Sen, J. Chem. Phys. 77, 4683 (1982).

22. W. Scheider, J. Phys. Chem. 79, 127 (1975).

23. J. B. Bates, J. C. Wang, and Y. T. Chu, Solid State Ionics 18-19, 1045 (1985).

24. R. de Levie, Electrochim. Acta 10, 113 (1965).

25. S. H. Liu, Phys. Rev. Lett. 55, 529 (1985).

26. A. Le Mehaute and G. Crepy, Solid State Ionics 9-10, 17 (1983).

27. L. Nyikos and T. Pajkossy, Electrochim. Acta 30, 1533 (1985).

28. T. C. Halsey, preprint (1986).

29. P.-G. de Gennes, C. R. Acad. Sc. Paris 295, 1061 (1982).

30. S. Alexander, in *Transport and Relaxation in Random Materials*, edited by M. Schlesinger (to be published).

List of Participants

A

Andreas Acrivos, Stanford University
Shlomo Alexander, Hebrew University
Tim Anderson, Union Oil Company of California
Michel Argaud, Elf Aquitaine
D. D. Awschalom, IBM–T. J. Watson Research Center

B

Jayanth Banavar, Schlumberger-Doll Research
Yves Bernabe, MIT
James Berryman, Lawrence Livermore Labs
Shobo Bhattacharya, Exxon Corporate Research
Daniel Bideau, Universite de Rennes
Kenneth Blum, Cornell University
Anthony Booer, Schlumberger Well Services
Ernest A. Boucher, University of Sussex
Richard Bradley, IBM–T. J. Watson Research Center
Alan Bray, Schlumberger-Doll Research
Steven Brown, Schlumberger-Doll Research

C

Russel Caflisch, Courant Institute–NYU
Moses Chan, Cornell University
Jing-Den Chen, Schlumberger-Doll Research
Philip Cheung, Schlumberger-Doll Research
N. I. Christensen, Purdue University
Christian Clavier, Schlumberger-Doll Research
George R. Coates, Schlumberger of Canada
Morrel Cohen, Exxon Corporate Research
Milton W. Cole, Pennsylvania State University
Winton Cornell, University of Rhode Island
Karl Coyner, New England Research
Stephen P. Cramer, Schlumberger-Doll Research
Charles Curtis, University of Sheffield

D

Anthony Roy Day, Michigan State University
J. A. de Waal, KSEPL
Emmanual Detournay, Dowell-Schlumberger
Madalena Dias, Schlumberger-Doll Research
Elizabeth Dussan, Schlumberger-Doll Research

E

Brian Evans, MIT

F

Shechao Feng, Schlumberger-Doll Research
Jan Finjord, Rogaland Regional College

G

Stephen Garoff, Schlumberger-Doll Research
Peter A. Goode, Schlumberger-Doll Research
Douglas H. Green, University of Wisconsin
J. Grolier, Elf Aquitaine
Robert Guyer, University of Massachusetts

H

Phillip Halleck, Schlumberger Perforating Center
Paul S. Hammond, Schlumberger-Doll Research
Siamak Hassanzadeh, Columbia University
Astley Hastings, Schlumberger-Doll Research
Ronald M. Haybron, Cleveland State University
Arch H. Heim, Schlumberger-Doll Research
James Helwig, Schlumberger-Doll Research
Mike Herron, Schlumberger-Doll Research
Susan Herron, Schlumberger-Doll Research
Stephen Hickman, MIT
Rune M. Holt, Petek-Sintef
Brian E. Hornby, Schlumberger-Doll Research
James Howard, Schlumberger-Doll Research

J

M. S. Jhon, Carnegie-Mellon University
David Johnson, Schlumberger-Doll Research

L

Leo P. Kadanoff, University of Chicago
William E. Kenyon, Schlumberger-Doll Research
Michael King, Sohio R&D
Joel Koplik, Schlumberger-Doll Research
Fikri Kuchuk, Schlumberger-Doll Research

L

Lynton Land, University of Texas
G. A. LaTorraca, Chevron Oil Field Research Company
B. Legait, Institut Francais du Petrole
Florian K. Lehner, Shell Research (KSEPL)
Roland Lenormand, Dowell-Schlumberger
Herbert Levine, Schlumberger-Doll Research
Max Lipsicas, Schlumberger-Doll Research
John Logan, Texas A&M University
Daniel Longeron, Institut Francais du Petrole
Jose Lopes, Schlumberger-Doll Research
Stefan Luthi, Schlumberger-Doll Research
J. W. Lyklema, DJV–TNO

M

M. H. Manghnani, University of Hawaii
C. McCann, The University of Reading
Kenneth Mendelson, Marquette University
Bob Metz, Carnegie-Mellon University
Michael J. Miksis, Northwestern University
Scott T. Milner, Exxon Corporate Research
Laurent Moinard, Schlumberger-Doll Research
F. D. Morgan, Texas A&M University
Sheena Murphy, Cornell University
William Murphy, Schlumberger-Doll Research

N

Alexandra Navrotsky, Princeton University
Andrew Norris, Rutgers University
Mark A. Novotny, Northwestern University

O

Gary Olhoeft, U.S. Geological Survey

P

Lindamae Peck, U.S. Army, Cold Regions Laboratory
P. Pfeifer, University of Missouri
Thomas J. Plona, Schlumberger-Doll Research
Richard A. Plumb, Schlumberger-Doll Research
James O. Pressley, Armstrong World Industries

R

T. S. Ramakrishnan, Schlumberger-Doll Research
Sidney Redner, Boston University
J. C. Roegiers, Dowell-Schlumberger
Mark Rosen, Grumman Corporate Research Center

S

Dominique Salin, Universite Pierre et Marie Curie
Ashok Sangani, Syracuse University
W. M. Saslow, Texas A&M University
Dale Schaefer, Sandia National Labs
Paul W. Schmidt, University of Missouri
Lawrence M. Schwartz, Schlumberger-Doll Research
David R. Scott, Caltech
Pabitra Sen, Schlumberger-Doll Research
Aaron Seriff, Shell Development Company
Yogeshwar Sharma, Schlumberger-Doll Research
Thomas Shaw, IBM–T. J. Watson Research Center
Alan Sibbit, Schlumberger-Doll Research
Arne Skjeltorp, Institute for Energy Technology

Duane Smith, Morgantown Energy Technology Center
Eve Sprunt, Mobil Research & Development
Christian Straley, Schlumberger-Doll Research

T

Yu-Chiung Teng, Columbia University
Michael Thambynayagam, Schlumberger-Doll Research
Pedro G. Toledo, University of Minnesota
Joseph P. Tomanic, Schlumberger-Doll Research
Micha Tomkiewicz, Brooklyn College, CUNY
Stephane Tyc, Harvard University

V

Wim Van Saarloos, AT&T Bell Labs
Dung Vo Thanh, Institut de Physique du Globe de Paris
D. U. von Rosenberg, Mobil Research and Development

W

Noboru Wada, Schlumberger-Doll Research
Joseph Walsh, MIT
H. F. Wang, University of Wisconsin
James Warnock, IBM–T. J. Watson Research Center
Bill Wepfer, Purdue University
Stanley Whittingham, Schlumberger-Doll Research
David J. Wilkinson, Schlumberger-Doll Research
Kenneth W. Winkler, Schlumberger-Doll Research
E. J. Witterholt, Standard Oil Production Company
Po-zen Wong, Schlumberger-Doll Research
J. R. Wood, Chevron Oil Field Research Company
R. W. Wunderlich, Chevron Oil Field Research

Y

Herbert H. Yuan, Shell Development Company

AIP Conference Proceedings

		L.C. Number	ISBN
No. 1	Feedback and Dynamic Control of Plasmas – 1970	70-141596	0-88318-100-2
No. 2	Particles and Fields – 1971 (Rochester)	71-184662	0-88318-101-0
No. 3	Thermal Expansion – 1971 (Corning)	72-76970	0-88318-102-9
No. 4	Superconductivity in d- and f-Band Metals (Rochester, 1971)	74-18879	0-88318-103-7
No. 5	Magnetism and Magnetic Materials – 1971 (2 parts) (Chicago)	59-2468	0-88318-104-5
No. 6	Particle Physics (Irvine, 1971)	72-81239	0-88318-105-3
No. 7	Exploring the History of Nuclear Physics – 1972	72-81883	0-88318-106-1
No. 8	Experimental Meson Spectroscopy –1972	72-88226	0-88318-107-X
No. 9	Cyclotrons – 1972 (Vancouver)	72-92798	0-88318-108-8
No. 10	Magnetism and Magnetic Materials – 1972	72-623469	0-88318-109-6
No. 11	Transport Phenomena – 1973 (Brown University Conference)	73-80682	0-88318-110-X
No. 12	Experiments on High Energy Particle Collisions – 1973 (Vanderbilt Conference)	73-81705	0-88318-111-8
No. 13	π-π Scattering – 1973 (Tallahassee Conference)	73-81704	0-88318-112-6
No. 14	Particles and Fields – 1973 (APS/DPF Berkeley)	73-91923	0-88318-113-4
No. 15	High Energy Collisions – 1973 (Stony Brook)	73-92324	0-88318-114-2
No. 16	Causality and Physical Theories (Wayne State University, 1973)	73-93420	0-88318-115-0
No. 17	Thermal Expansion – 1973 (Lake of the Ozarks)	73-94415	0-88318-116-9
No. 18	Magnetism and Magnetic Materials – 1973 (2 parts) (Boston)	59-2468	0-88318-117-7
No. 19	Physics and the Energy Problem – 1974 (APS Chicago)	73-94416	0-88318-118-5
No. 20	Tetrahedrally Bonded Amorphous Semiconductors (Yorktown Heights, 1974)	74-80145	0-88318-119-3
No. 21	Experimental Meson Spectroscopy – 1974 (Boston)	74-82628	0-88318-120-7
No. 22	Neutrinos – 1974 (Philadelphia)	74-82413	0-88318-121-5
No. 23	Particles and Fields – 1974 (APS/DPF Williamsburg)	74-27575	0-88318-122-3
No. 24	Magnetism and Magnetic Materials – 1974 (20th Annual Conference, San Francisco)	75-2647	0-88318-123-1
No. 25	Efficient Use of Energy (The APS Studies on the Technical Aspects of the More Efficient Use of Energy)	75-18227	0-88318-124-X
No. 26	High-Energy Physics and Nuclear Structure – 1975 (Santa Fe and Los Alamos)	75-26411	0-88318-125-8
No. 27	Topics in Statistical Mechanics and Biophysics: A Memorial to Julius L. Jackson (Wayne State University, 1975)	75-36309	0-88318-126-6
No. 28	Physics and Our World: A Symposium in Honor of Victor F. Weisskopf (M.I.T., 1974)	76-7207	0-88318-127-4

No. 29	Magnetism and Magnetic Materials – 1975 (21st Annual Conference, Philadelphia)	76-10931	0-88318-128-2
No. 30	Particle Searches and Discoveries – 1976 (Vanderbilt Conference)	76-19949	0-88318-129-0
No. 31	Structure and Excitations of Amorphous Solids (Williamsburg, VA, 1976)	76-22279	0-88318-130-4
No. 32	Materials Technology – 1976 (APS New York Meeting)	76-27967	0-88318-131-2
No. 33	Meson-Nuclear Physics – 1976 (Carnegie-Mellon Conference)	76-26811	0-88318-132-0
No. 34	Magnetism and Magnetic Materials – 1976 (Joint MMM-Intermag Conference, Pittsburgh)	76-47106	0-88318-133-9
No. 35	High Energy Physics with Polarized Beams and Targets (Argonne, 1976)	76-50181	0-88318-134-7
No. 36	Momentum Wave Functions – 1976 (Indiana University)	77-82145	0-88318-135-5
No. 37	Weak Interaction Physics – 1977 (Indiana University)	77-83344	0-88318-136-3
No. 38	Workshop on New Directions in Mossbauer Spectroscopy (Argonne, 1977)	77-90635	0-88318-137-1
No. 39	Physics Careers, Employment and Education (Penn State, 1977)	77-94053	0-88318-138-X
No. 40	Electrical Transport and Optical Properties of Inhomogeneous Media (Ohio State University, 1977)	78-54319	0-88318-139-8
No. 41	Nucleon-Nucleon Interactions – 1977 (Vancouver)	78-54249	0-88318-140-1
No. 42	Higher Energy Polarized Proton Beams (Ann Arbor, 1977)	78-55682	0-88318-141-X
No. 43	Particles and Fields – 1977 (APS/DPF, Argonne)	78-55683	0-88318-142-8
No. 44	Future Trends in Superconductive Electronics (Charlottesville, 1978)	77-9240	0-88318-143-6
No. 45	New Results in High Energy Physics – 1978 (Vanderbilt Conference)	78-67196	0-88318-144-4
No. 46	Topics in Nonlinear Dynamics (La Jolla Institute)	78-57870	0-88318-145-2
No. 47	Clustering Aspects of Nuclear Structure and Nuclear Reactions (Winnepeg, 1978)	78-64942	0-88318-146-0
No. 48	Current Trends in the Theory of Fields (Tallahassee, 1978)	78-72948	0-88318-147-9
No. 49	Cosmic Rays and Particle Physics – 1978 (Bartol Conference)	79-50489	0-88318-148-7
No. 50	Laser-Solid Interactions and Laser Processing – 1978 (Boston)	79-51564	0-88318-149-5
No. 51	High Energy Physics with Polarized Beams and Polarized Targets (Argonne, 1978)	79-64565	0-88318-150-9
No. 52	Long-Distance Neutrino Detection – 1978 (C.L. Cowan Memorial Symposium)	79-52078	0-88318-151-7
No. 53	Modulated Structures – 1979 (Kailua Kona, Hawaii)	79-53846	0-88318-152-5
No. 54	Meson-Nuclear Physics – 1979 (Houston)	79-53978	0-88318-153-3
No. 55	Quantum Chromodynamics (La Jolla, 1978)	79-54969	0-88318-154-1
No. 56	Particle Acceleration Mechanisms in Astrophysics (La Jolla, 1979)	79-55844	0-88318-155-X

No. 57	Nonlinear Dynamics and the Beam-Beam Interaction (Brookhaven, 1979)	79-57341	0-88318-156-8
No. 58	Inhomogeneous Superconductors – 1979 (Berkeley Springs, W.V.)	79-57620	0-88318-157-6
No. 59	Particles and Fields – 1979 (APS/DPF Montreal)	80-66631	0-88318-158-4
No. 60	History of the ZGS (Argonne, 1979)	80-67694	0-88318-159-2
No. 61	Aspects of the Kinetics and Dynamics of Surface Reactions (La Jolla Institute, 1979)	80-68004	0-88318-160-6
No. 62	High Energy e^+e^- Interactions (Vanderbilt, 1980)	80-53377	0-88318-161-4
No. 63	Supernovae Spectra (La Jolla, 1980)	80-70019	0-88318-162-2
No. 64	Laboratory EXAFS Facilities – 1980 (Univ. of Washington)	80-70579	0-88318-163-0
No. 65	Optics in Four Dimensions – 1980 (ICO, Ensenada)	80-70771	0-88318-164-9
No. 66	Physics in the Automotive Industry – 1980 (APS/AAPT Topical Conference)	80-70987	0-88318-165-7
No. 67	Experimental Meson Spectroscopy – 1980 (Sixth International Conference, Brookhaven)	80-71123	0-88318-166-5
No. 68	High Energy Physics – 1980 (XX International Conference, Madison)	81-65032	0-88318-167-3
No. 69	Polarization Phenomena in Nuclear Physics – 1980 (Fifth International Symposium, Santa Fe)	81-65107	0-88318-168-1
No. 70	Chemistry and Physics of Coal Utilization – 1980 (APS, Morgantown)	81-65106	0-88318-169-X
No. 71	Group Theory and its Applications in Physics – 1980 (Latin American School of Physics, Mexico City)	81-66132	0-88318-170-3
No. 72	Weak Interactions as a Probe of Unification (Virginia Polytechnic Institute – 1980)	81-67184	0-88318-171-1
No. 73	Tetrahedrally Bonded Amorphous Semiconductors (Carefree, Arizona, 1981)	81-67419	0-88318-172-X
No. 74	Perturbative Quantum Chromodynamics (Tallahassee, 1981)	81-70372	0-88318-173-8
No. 75	Low Energy X-Ray Diagnostics – 1981 (Monterey)	81-69841	0-88318-174-6
No. 76	Nonlinear Properties of Internal Waves (La Jolla Institute, 1981)	81-71062	0-88318-175-4
No. 77	Gamma Ray Transients and Related Astrophysical Phenomena (La Jolla Institute, 1981)	81-71543	0-88318-176-2
No. 78	Shock Waves in Condensed Matter – 1981 (Menlo Park)	82-70014	0-88318-177-0
No. 79	Pion Production and Absorption in Nuclei – 1981 (Indiana University Cyclotron Facility)	82-70678	0-88318-178-9
No. 80	Polarized Proton Ion Sources (Ann Arbor, 1981)	82-71025	0-88318-179-7
No. 81	Particles and Fields –1981: Testing the Standard Model (APS/DPF, Santa Cruz)	82-71156	0-88318-180-0
No. 82	Interpretation of Climate and Photochemical Models, Ozone and Temperature Measurements (La Jolla Institute, 1981)	82-71345	0-88318-181-9
No. 83	The Galactic Center (Cal. Inst. of Tech., 1982)	82-71635	0-88318-182-7

No. 84	Physics in the Steel Industry (APS/AISI, Lehigh University, 1981)	82-72033	0-88318-183-5
No. 85	Proton-Antiproton Collider Physics –1981 (Madison, Wisconsin)	82-72141	0-88318-184-3
No. 86	Momentum Wave Functions – 1982 (Adelaide, Australia)	82-72375	0-88318-185-1
No. 87	Physics of High Energy Particle Accelerators (Fermilab Summer School, 1981)	82-72421	0-88318-186-X
No. 88	Mathematical Methods in Hydrodynamics and Integrability in Dynamical Systems (La Jolla Institute, 1981)	82-72462	0-88318-187-8
No. 89	Neutron Scattering – 1981 (Argonne National Laboratory)	82-73094	0-88318-188-6
No. 90	Laser Techniques for Extreme Ultraviolt Spectroscopy (Boulder, 1982)	82-73205	0-88318-189-4
No. 91	Laser Acceleration of Particles (Los Alamos, 1982)	82-73361	0-88318-190-8
No. 92	The State of Particle Accelerators and High Energy Physics (Fermilab, 1981)	82-73861	0-88318-191-6
No. 93	Novel Results in Particle Physics (Vanderbilt, 1982)	82-73954	0-88318-192-4
No. 94	X-Ray and Atomic Inner-Shell Physics – 1982 (International Conference, U. of Oregon)	82-74075	0-88318-193-2
No. 95	High Energy Spin Physics – 1982 (Brookhaven National Laboratory)	83-70154	0-88318-194-0
No. 96	Science Underground (Los Alamos, 1982)	83-70377	0-88318-195-9
No. 97	The Interaction Between Medium Energy Nucleons in Nuclei – 1982 (Indiana University)	83-70649	0-88318-196-7
No. 98	Particles and Fields – 1982 (APS/DPF University of Maryland)	83-70807	0-88318-197-5
No. 99	Neutrino Mass and Gauge Structure of Weak Interactions (Telemark, 1982)	83-71072	0-88318-198-3
No. 100	Excimer Lasers – 1983 (OSA, Lake Tahoe, Nevada)	83-71437	0-88318-199-1
No. 101	Positron-Electron Pairs in Astrophysics (Goddard Space Flight Center, 1983)	83-71926	0-88318-200-9
No. 102	Intense Medium Energy Sources of Strangeness (UC-Sant Cruz, 1983)	83-72261	0-88318-201-7
No. 103	Quantum Fluids and Solids – 1983 (Sanibel Island, Florida)	83-72440	0-88318-202-5
No. 104	Physics, Technology and the Nuclear Arms Race (APS Baltimore –1983)	83-72533	0-88318-203-3
No. 105	Physics of High Energy Particle Accelerators (SLAC Summer School, 1982)	83-72986	0-88318-304-8
No. 106	Predictability of Fluid Motions (La Jolla Institute, 1983)	83-73641	0-88318-305-6
No. 107	Physics and Chemistry of Porous Media (Schlumberger-Doll Research, 1983)	83-73640	0-88318-306-4
No. 108	The Time Projection Chamber (TRIUMF, Vancouver, 1983)	83-83445	0-88318-307-2

No. 109	Random Walks and Their Applications in the Physical and Biological Sciences (NBS/La Jolla Institute, 1982)	84-70208	0-88318-308-0
No. 110	Hadron Substructure in Nuclear Physics (Indiana University, 1983)	84-70165	0-88318-309-9
No. 111	Production and Neutralization of Negative Ions and Beams (3rd Int'l Symposium, Brookhaven, 1983)	84-70379	0-88318-310-2
No. 112	Particles and Fields – 1983 (APS/DPF, Blacksburg, VA)	84-70378	0-88318-311-0
No. 113	Experimental Meson Spectroscopy – 1983 (Seventh International Conference, Brookhaven)	84-70910	0-88318-312-9
No. 114	Low Energy Tests of Conservation Laws in Particle Physics (Blacksburg, VA, 1983)	84-71157	0-88318-313-7
No. 115	High Energy Transients in Astrophysics (Santa Cruz, CA, 1983)	84-71205	0-88318-314-5
No. 116	Problems in Unification and Supergravity (La Jolla Institute, 1983)	84-71246	0-88318-315-3
No. 117	Polarized Proton Ion Sources (TRIUMF, Vancouver, 1983)	84-71235	0-88318-316-1
No. 118	Free Electron Generation of Extreme Ultraviolet Coherent Radiation (Brookhaven/OSA, 1983)	84-71539	0-88318-317-X
No. 119	Laser Techniques in the Extreme Ultraviolet (OSA, Boulder, Colorado, 1984)	84-72128	0-88318-318-8
No. 120	Optical Effects in Amorphous Semiconductors (Snowbird, Utah, 1984)	84-72419	0-88318-319-6
No. 121	High Energy e^+e^- Interactions (Vanderbilt, 1984)	84-72632	0-88318-320-X
No. 122	The Physics of VLSI (Xerox, Palo Alto, 1984)	84-72729	0-88318-321-8
No. 123	Intersections Between Particle and Nuclear Physics (Steamboat Springs, 1984)	84-72790	0-88318-322-6
No. 124	Neutron-Nucleus Collisions – A Probe of Nuclear Structure (Burr Oak State Park - 1984)	84-73216	0-88318-323-4
No. 125	Capture Gamma-Ray Spectroscopy and Related Topics – 1984 (Internat. Symposium, Knoxville)	84-73303	0-88318-324-2
No. 126	Solar Neutrinos and Neutrino Astronomy (Homestake, 1984)	84-63143	0-88318-325-0
No. 127	Physics of High Energy Particle Accelerators (BNL/SUNY Summer School, 1983)	85-70057	0-88318-326-9
No. 128	Nuclear Physics with Stored, Cooled Beams (McCormick's Creek State Park, Indiana, 1984)	85-71167	0-88318-327-7
No. 129	Radiofrequency Plasma Heating (Sixth Topical Conference, Callaway Gardens, GA, 1985)	85-48027	0-88318-328-5
No. 130	Laser Acceleration of Particles (Malibu, California, 1985)	85-48028	0-88318-329-3
No. 131	Workshop on Polarized ^{3}He Beams and Targets (Princeton, New Jersey, 1984)	85-48026	0-88318-330-7
No. 132	Hadron Spectroscopy–1985 (International Conference, Univ. of Maryland)	85-72537	0-88318-331-5

No. 133	Hadronic Probes and Nuclear Interactions (Arizona State University, 1985)	85-72638	0-88318-332-3
No. 134	The State of High Energy Physics (BNL/SUNY Summer School, 1983)	85-73170	0-88318-333-1
No. 135	Energy Sources: Conservation and Renewables (APS, Washington, DC, 1985)	85-73019	0-88318-334-X
No. 136	Atomic Theory Workshop on Relativistic and QED Effects in Heavy Atoms	85-73790	0-88318-335-8
No. 137	Polymer-Flow Interaction (La Jolla Institute, 1985)	85-73915	0-88318-336-6
No. 138	Frontiers in Electronic Materials and Processing (Houston, TX, 1985)	86-70108	0-88318-337-4
No. 139	High-Current, High-Brightness, and High-Duty Factor Ion Injectors (La Jolla Institute, 1985)	86-70245	0-88318-338-2
No. 140	Boron-Rich Solids (Albuquerque, NM, 1985)	86-70246	0-88318-339-0
No. 141	Gamma-Ray Bursts (Stanford, CA, 1984)	86-70761	0-88318-340-4
No. 142	Nuclear Structure at High Spin, Excitation, and Momentum Transfer (Indiana University, 1985)	86-70837	0-88318-341-2
No. 143	Mexican School of Particles and Fields (Oaxtepec, México, 1984)	86-81187	0-88318-342-0
No. 144	Magnetospheric Phenomena in Astrophysics (Los Alamos, 1984)	86-71149	0-88318-343-9
No. 145	Polarized Beams at SSC & Polarized Antiprotons (Ann Arbor, MI & Bodega Bay, CA, 1985)	86-71343	0-88318-344-7
No. 146	Advances in Laser Science–I (Dallas, TX, 1985)	86-71536	0-88318-345-5
No. 147	Short Wavelength Coherent Radiation: Generation and Applications (Monterey, CA, 1986)	86-71674	0-88318-346-3
No. 148	Space Colonization: Technology and The Liberal Arts (Geneva, NY, 1985)	86-71675	0-88318-347-1
No. 149	Physics and Chemistry of Protective Coatings (Universal City, CA, 1985)	86-72019	0-88318-348-X
No. 150	Intersections Between Particle and Nuclear Physics (Lake Louise, Canada, 1986)	86-72018	0-88318-349-8
No. 151	Neural Networks for Computing (Snowbird, UT, 1986)	86-72481	0-88318-351-X
No. 152	Heavy Ion Inertial Fusion (Washington, DC, 1986)	86-73185	0-88318-352-8
No. 153	Physics of Particle Accelerators (SLAC Summer School, 1985) (Fermilab Summer School, 1984)	87-70103	0-88318-353-6